Sidney I. (Sidney Irving) Smith

Preliminary Report on the Brachyura and Anomura dredged in deep Water off the South Coast of New England by the United States Fish Commission in 1880, 1881 and 1882

Sidney I. (Sidney Irving) Smith

Preliminary Report on the Brachyura and Anomura dredged in deep Water off the South Coast of New England by the United States Fish Commission in 1880, 1881 and 1882

ISBN/EAN: 9783741184413

Manufactured in Europe, USA, Canada, Australia, Japa

Cover: Foto ©berggeist007 / pixelio.de

Manufactured and distributed by brebook publishing software (www.brebook.com)

Sidney I. (Sidney Irving) Smith

Preliminary Report on the Brachyura and Anomura dredged in deep Water off the South Coast of New England by the United States Fish Commission in 1880, 1881 and 1882

PROCEEDINGS

OF THE

UNITED STATES NATIONAL MUSEUM.

1883.

Vol. VI, No. 1.　　Washington, D. C.　　June 18, 1883.

PRELIMINARY REPORT ON THE BRACHYURA AND ANOMURA DREDGED IN DEEP WATER OFF THE SOUTH COAST OF NEW ENGLAND BY THE UNITED STATES FISH COMMISSION IN 1880, 1881, AND 1882.

By SIDNEY I. SMITH.

This report is intended to be supplementary to my preliminary notice of the crustacea dredged in the same region in 1880 (these Proceedings, iii, pp. 413–452), and to include all the species of Brachyura and Anomura obtained off Martha's Vineyard at depths greater than 50 fathoms. The crustacea dredged off the mouth of Chesapeake Bay in 1880, and off the capes of the Delaware in 1881, will be the subject of a separate report, but the specimens from these dredgings are included in the following lists of specimens examined as far as the species to which they belong are contained in the present report.

A few of the species described as new in the preliminary notice above referred to were almost simultaneously described by A. Milne-Edwards in one of the reports of the Blake dredgings, under the supervision of Alexander Agassiz, in 1877, 1878, 1879 (Bull. Mus. Comp. Zool. Cambridge, vol. viii, No. 1, December 29, 1880), of which I had no knowledge whatever until after my paper was printed (January, 1881), and which was not published until after my last proof had been returned to the Public Printer (December 24, 1880). I have had much difficulty in identifying Milne-Edwards's species, but have adopted his names wherever it was possible to recognize his species. In determining some of these species I have been greatly aided by the kindness of Prof. Walter Faxon, who has sent me for examination some of the type specimens in the Museum of Comparative Zoology.[*]

The last season's dredging off Martha's Vineyard reveals the total, or almost total, disappearance of several of the larger species of crustacea which were exceedingly abundant in the same region in 1880 and 1881. The most remarkable cases are those of *Euprognatha rastellifera*, *Collodes robustus*, *Catapagurus Sharreri*, *Munida Caribæa?* Smith, and

---

[*] While the manuscript of this report was in the hands of the printer, the following work of Milne-Edwards was received: Recueil de figures de Crustacés nouveaux ou peu connus. 1ère livraison. April, 1883. A considerable number of Milne-Edw..'s new species are provisionally figured in this work, but it does not seem to make any changes in the proofs of the following pages necessary, except under *Anoplonotus politus*, which was doubtfully referred to *Elasmonotus* in the original manuscript, but for which the new generic name has been inserted in the proof.—May 29, 1883.

Pontophilus brevirostris, all of which were exceedingly abundant in 1880 and 1881; but of the first two not a specimen was taken the past season, of the *Munida* only a single specimen, and that on the last trip, and of the other species only a very few specimens. *Lambrus Verrillii*, *Acanthocarpus Alexandri*, *Latreillia elegans*, *Homola barbata*, and *Anoplonotus politus*, which were each taken several times in 1880 and 1881, were none of them taken in 1882; they were far less abundant than the other species, however, and the non-occurrence of some of them was very likely accidental; but the disappearance of part of them at least was undoubtedly due to the same causes which occasioned the disappearance of the more abundant species. The disappearance of these species was undoubtedly connected directly with the similar disappearance of the tile-fish (*Lopholatilus*) from the same region, and on this account specially I give in detail, for many of the species enumerated beyond, the tables of specimens examined from the region explored by the Fish Commission; and to these I have usually added the specimens which I have examined from the collection made by Alexander Agassiz on the Blake in 1880. All the species mentioned above as having disappeared in 1882 were specially characteristic of the narrow belt of comparatively warm water (approximately 50° F.), in from 60 to 100 fathoms, which has a more southern fauna than the colder waters either side. Professor Verrill has suggested (Amer. Jour. Sci., III, xxiv, p. 366, 1882) that there was a great destruction of life in this belt, caused by a severe storm, in the winter of 1881–'82, which agitated the bottom-water and forced outward the cold water that even in summer occupies the great area of shallow sea along the coast, thus causing a sudden lowering of the temperature along the warmer belt inhabited by the tile-fish and the crustacea referred to.

In the following tables of specimens examined the latitude and longitude, depth, nature of bottom, &c., are copied from the list of dredging sfations of the Fish Commission for 1880, 1881, and 1882, in the Bulletin of the Fish Commission, vol. ii, pp. 119 to 131, 1882, where further details in regard to temperature, &c., are given. In indicating the nature of the bottom, the Coast Survey system of abbreviations is used. In the column for the number of specimens examined, *l* is used to indicate large specimens; *s*, small specimens; and *y*, young. When the sexes were not counted separately the whole number of specimens examined is placed in the middle of the column; when the sexes were counted separately the number of males is put on the right, the number of females on the left, and the number of young in the middle, followed by the letter *y*. As a basis for ascertaining the breeding season, I have, in a great number of cases, noted the presence or absence of egg-bearing females; when the number of such females was counted it is entered in the appropriate column; when specimens carrying eggs were found, but not counted, a plus sign, +, is used; and when none of the specimens examined were carrying eggs a zero, 0, is used.

BRACHYURA.

MAIOIDEA.

Amathia Agassizii Smith, Bull. Mus. Comp. Zool., Cambridge, x, p. 1, pl. 2, figs. 2, 3, 1882.

Specimens examined.

Station No.	Locality.		Depth in fathoms.	Nature of bottom.	When collected.	No. of specimens.		
	N. lat.	W. long.				♂	♀	With eggs.
	OFF MARTHA'S VINEYARD.							
	° ′ ″	° ′ ″			1881.			
939	39 53 00	69 50 30	264	gn. S. M.	Aug. 4	1		
1032	39 56 00	69 22 00	206	yl. M.	Sept. 14	1	1 y.	
					1882.			
1113	39 57 00	70 37 00	192	gn. M.	Aug. 22	1		
1154	39 55 31	70 39 00	193	S. M.	Oct. 4	3		

In the original description above referred to it is stated that this species resembles *Amathia Carpenteri* Norman (*Scyramathia Carpenteri* A. M.-Edwards); it is, however, probably not closely allied or even congeneric with that species, but apparently closely allied to *Amathia crassa* A. M.-Edwards, and possibly identical with it. I was misled in regard to the armament of the carapax of *Scyramathia Carpenteri* by the woodcut given in the Depths of the Sea (no description of the species has yet appeared), for Milne-Edwards states that the species is closely allied to *Scyra umbonata* Stimpson, certainly a very different species from *Amathia Agassizii*, and has united them in his new genus *Scyramathia*.

As indicated above, all the specimens seen are males. One of these is much larger than the larger of the two original specimens described and figured in my report above referred to, but differs very little from it, although the spines of the horizontal series on the branchial region, above the bases of the cheliped and first ambulatory leg, are considerably longer, and there are two well-developed spines, instead of two or three small ones, on the lateral margin back of the anterior angle of the buccal area. Measurements of this specimen are given in the last column of the accompanying table of measurements. The other specimens show all gradations between this and the young specimens originally described.

Measurements in millimeters.

	Station—			
	1032.	939.	1113.	1154.
Sex	Young.	♂	♂	♂
Length of carapax, including rostral and posterior spines	15.5	21.5	26.3	53.5
Length of carapax from base of rostral to tip of posterior spines	10.0	14.0	18.0	42.5
Length of carapax, excluding rostral and posterior spines	9.1	12.1	16.5	40.5
Length of rostral horns or spines	5.7	8.0	9.0	12.0
Breadth of carapax, including lateral spines	11.5	16.0	19.5	39.5
Breadth of carapax, excluding lateral spines	6.8	9.0	12.3	32.0
Length of branchial spine	3.1	4.8	5.2	6.0
Length of cheliped	11.0	15.0	21.5	62.0
Length of chela	4.6	6.4	10.0	28.5
Breadth of chela	0.9	1.3	1.6	4.0
Length of dactylus	2.0	2.5	3.8	5.4
Length of first ambulatory leg	18.0	25.0	37.0	....
Length of dactylus	3.5	4.0	6.7	....
Length of second ambulatory leg	15.0	22.5	31.5	95.0
Length of dactylus	3.2	4.8	6.0	17.5

Amathia Tanneri, sp. nov.

Allied to the last species, but readily distinguished from it by the narrower carapax with longer and less diverging rostral horns and fewer and more nearly equal spines, and by having a single spine only on the base of the antenna.

Male.—The carapax, excluding the rostral horns and the spines, is about as broad as long. The rostral horns are nearly straight, much less divergent than in *A. Agassizii,* and, in the larger of the two specimens seen, much more than half as long as the rest of the carapax. The supraorbital spine and the postorbital process are as in *A. Agassizii,* but the basal segment of the antenna is unarmed except by a single spine at the distal end. There are four long and approximately equidistant spines on the mesial line of the carapax, the two anterior on the gastric region and smaller than the others, which are on the cardiac region, the posterior being near the posterior margin and projecting slightly backward over it. There are no prominent spines on the gastric region except the two median, but there is a minute tubercle or rudimentary spine either side about equidistant from the two median, and on the cardiac region there are no spines or tubercles whatever except the two median. There is a single long hepatic and a great branchial spine, as in *A. Agassizii,* but there are no other spines or tubercles on the branchial region except two, about as long as the cardiac spines, and about equidistant from each other and from the great branchial, the posterior gastric, and the anterior cardiac. The anterior angle of the buccal area projects in a dentiform process either side, as in *A. Agassizii,* and back of this the prominent margin of the pleural region is armed with three small tubercles or rudimentary spines. There are no spines or tubercles on the side of the branchial region above the basis of the cheliped and first ambulatory leg, and no tubercles whatever on the postero-lateral margins.

The chelipeds and ambulatory legs are essentially as in *A. Agassizii.*

The number and arrangement of the dorsal spines of the carapax appear to be nearly as in *A. hystrix* Stimpson, as figured by A. Milne-Edwards (Crust. Région Mexicaine, p. 134, pl. 28, fig. 1, 1878), except that the lateral spines of the gastric region are obsolete in *A. Tanneri;* but all the spines are very greatly longer in *hystrix*, which appears to be a very distinct species.

Measurements in millimeters.

	Station—	
	1038.	1043.
Sex	♂	♂
Length of carapax, including rostral and posterior spines	15+	28.0
Length of carapax from base of rostral to tip of posterior spines	11.0	18.0
Length of carapax, excluding rostral and posterior spines	10.2	16.2
Length of rostral horns or spines	4+	11.0
Breadth of carapax, including lateral spines	10.0	18.0
Breadth of carapax, excluding lateral spines	6.7	11.2
Length of branchial spine	1.9	4.0
Length of cheliped	12.0	20.0
Length of chela	5.5	9.0
Breadth of chela	1.1	1.8
Length of dactylus	2.0	3.5
Length of first ambulatory leg	21.0	35.0
Length of dactylus	3.7	6.2
Length of second ambulatory leg	16.0	27.0
Length of dactylus	3.0	5.9

Specimens examined.

Station No	Locality.		Depth in fathoms.	Nature of bottom.	When collected.	No. of specimens.	
	N. lat.	W. long.				♂	♀
	OFF MARTHA'S VINEYARD.						
	° ′	° ′			1881.		
1038	39 58	70 06	146	S. Sh.	Sept. 21	1 y.	
	OFF DELAWARE BAY.						
	° ′	° ′			1881.		
1043	38 39	73 11	130	S.	Oct. 10	1	

Hyas coarctatus Leach.

Taken at a number of stations off Martha's Vineyard, in 86 to 158 fathoms, and also in much shallower water near Block Island and off No Man's Land. Four male specimens were taken off Chesapeake Bay in 1880, station 900, N. lat. 37° 19′, W. long. 74° 41′, 31 fath., sand—the farthest south the species has been noticed.

Collodes robustus, sp. nov.

Collodes depressus Smith, Proc. National Mus., iii, p. 414, 1881 (*non* A. M.-Edwards.)

A careful examination of one of the type specimens of *C. depressus* convinces me that the specimens which I have referred to that species are really a distinct but closely allied and much larger species. Very small specimens, 10ᵐᵐ or less in length of carapax, resemble the *depressus* very much, but are distinguished from Milne-Edwards's figures and the type specimen referred to by the less regularly triangular outline of the carapax, the hepatic and branchial regions being much more protuberant; by the acute rostral horns, more widely separated at their tips; by the much longer interantennular spine, which is fully as long as in *Euprognatha rastellifera*; by the short and conical or even tuberculiform gastric and cardiac spines; by the spine of the first somite of the abdomen being directed backward instead of upward; and by the more slender chelæ.

Male.—In large males over 20ᵐᵐ in length of carapax, the carapax is a little over three-fourths as broad as long, and thickly covered, as well as nearly all other parts of the animal except the chelæ, with strongly curved hairs or setæ, which, in every specimen seen, persistantly retain a thick coating of soft mud. The rostral horns are slender and separated by a rounded sinus, at the bottom of which the interantennular spine, or true rostrum, which is much longer than the rostral horns and grooved longitudinally in front, projects downward and about as far forward as the rostral horns. The basal segment of the antenna is armed with a lateral and an inferior ridge, each divided into three to five short spiniform teeth. The postorbital processes are broad, but acutely triangular, and project as far as the tips of the eyes. The dorsal surface is thickly covered with granular tubercles, and there is a slight tubercular elevation, but little more prominent than the tubercles of the general surface, on the gastric region, and another on the cardiac, in place of the spines in the young. The hepatic region is divided obliquely near the middle by a deep sulcus into two lobes, of which the superior projects in a rounded prominence, which is very conspicuous as seen from above, while the inferior is crossed longitudinally by the pleurotergal suture and below it armed with a short series of small tuberculiform spines. The branchial regions are prominent, swollen, and evenly tuberculated.

The chelipeds are stout and approximately once and a half as long as the carapax; the merus is triquetral with the angles armed more or less with tubercles or tuberculiform spines; the whole outer surface of the carpus is similarly armed. The chela is approximately two-thirds as long as the carapax, naked, smooth, polished, and unarmed, except a very few tubercles on the inner surface and near the proximal ends of the upper and under edges; the body is nearly as long as the digits, thick and swollen; and the digits are compressed, somewhat grooved longitudinally, very slightly curved, gaping at the bases, and with the prehensile edges slightly and irregularly crenate. The ambulatory legs are hairy to very near the tips, but are otherwise unarmed and smooth throughout, and all the segments are subcylindrical; the first are about two and a half times as long as the carapax, the others successively shorter, and the last considerably less than twice as long as the carapax; the dactyli are considerably curved, slender, and tapered near the acute chitinous tips.

The sternum is tuberculose, like the dorsal surface of the carapax, except upon the concave portion between the bases of the chelipeds, where it is smooth.

The first somite of the abdomen is tuberculose, like the carapax, and armed with a low tuberculiform prominence, in place of the spine in the young. The second somite is very short and scarcely wider than the first. The third is widest of all, and from it the abdomen is regularly narrowed to the seventh somite, which is anchylosed with the sixth, as in *Euprognatha rastellifera*, triangular, with the tip obtuse, and nearly as broad as long.

Female.—The females appear not to attain the adult sexual characters until the carapax is about 12^{mm} in length, apparently never attain as great size as the males, and as usual resemble the young, although they lose the gastric, cardiac, and abdominal spines fully as early as the males. The carapax is slightly more convex and the branchial regions somewhat less swollen than in the male. The chelipeds remain small and weak, the chelæ slender as in the young, and the ambulatory legs proportionally shorter than in the male.

The proportions of the carapax, chelipeds, and ambulatory legs in the young and adults of both sexes are well shown by the accompanying table of measurements.

Measurements in millimeters and hundredths of length of carapax.

	Station—									
	874	949	940	940	040	040	873	1000	950	940
Sex	♂*	♂	♂	♂	♂	♂	♀*	♀*	♀	♀
Length of carapax, including frontal teeth	9.7	12.5	14.8	23.3	25.7	27.0	8.2	10.7	13.3	18.3
Greatest breadth of carapax	6.6	9.1	10.8	17.9	20.7	21.2	5.8	7.0	10.3	14.1
Same in hundredths of length of carapax	68	73	73	77	77	78	71	68	70	77
Length of cheliped	11.0	16.0	18.0	34.0	38.0	40.0	9.5	11.0	14.0	19.0
Length of chela	4.8	6.2	7.8	14.8	17.6	18.5	3.5	4.5	5.5	7.9
Same in hundredths of length of carapax	49	50	53	64	66	68	41	42	42	43
Height of chela	1.2	2.1	2.7	6.6	7.5	8.0	0.9	1.1	1.4	2.1
Same in hundredths of length of carapax	12	17	18	28	29	30	11	10	11	12
Length of dactylus	2.7	3.6	4.4	8.0	9.1	10.2	2.1	2.6	3.2	4.7
Length of first ambulatory leg	20.0	28.0	34.0	58.0	65.0	68.0	15.0	19.0	24.0	32.0
Length of propodus	5.0	6.5	8.5	14.5	15.5	17.0	3.1	4.1	6.0	7.5
Length of dactylus	3.9	5.4	7.0	12.0	13.4	14.5	2.7	3.9	5.2	7.0
Length of fourth ambulatory leg	17.0	22.0	27.0	42.0	46.0	47.0	......	15.5	20.0	25.0
Length of propodus	3.8	5.0	6.9	10.1	12.0	12.6	......	2.9	4.8	6.7
Length of dactylus	3.8	5.0	6.3	9.2	10.0	10.2	......	2.8	4.7	6.2

* Immature individuals.

The number and arrangement of the branchiæ are the same as in *Euprognatha rastellifera*, but there are well-developed epipods on all three pairs of maxillipeds, those on the second being narrow, but as long as the merus of the endopod, so that the formula is:

	Somite—								
	VII.	VIII.	IX.	X.	XI.	XII.	XIII.	XIV.	Total.
Epipods	1	1	1	0	0	0	0	0	(3)
Podobranchiæ	0	1	1	0	0	0	0	0	2
Arthrobranchiæ	0	0	0	2	0	0	0	0	2
Pleurobranchiæ	0	0	0	2	1	1	0	0	4
									8+(3)

Specimens examined.

Station No.	Locality.		Depth in fathoms.	Nature of bottom.	When collected.	No. of specimens.		
	N. lat.	W. long.				♂	♀	With eggs.
	OFF MARTHA'S VINEYARD.							
	° ′ ″	° ′ ″			1880.			
865	40 05 00	70 23 00	65	fne. S. M.	Sept. 4	1	1	1
871	40 02 54	70 23 40	115	fne. S. M.	Sept. 4	8	1	0
872	40 05 39	70 23 52	86	S. G. Sh. sponges.	Sept. 4	1		
873	40 02 00	70 57 00	100	sft. M.	Sept. 13	3	3	0
874	40 00 00	70 57 00	85	sft. M.	Sept. 13	4	4	0
875	39 57 00	70 57 30	120	sft. M.	Sept. 13		1	0
878	39 55 00	70 54 15	142	sft. M.	Sept. 13	1	1	1
					1881.			
921	40 07 48	70 43 54	67	gn. M.	July 16	4	1	0
922	40 03 48	70 45 54	71	gn. M. S.	July 16	2		
940	39 54 00	69 51 30	134	hrd. S. sponges.	Aug. 4	34	9	5
941	40 01 00	69 56 00	79	hrd. S. M.	Aug. 4	8	9	9
949	40 03 00	70 31 00	100	yl. M.	Aug. 23	5	5	5
950	40 07 00	70 32 00	71	S. Sh. M.	Aug. 23	3	2	2
1036	39 58 00	69 30 00	94	S.	Sept. 14	2	1	0
1038	39 58 00	70 06 00	146	S. Sh.	Sept. 21	1		
1040	40 09 00	70 06 00	93	S. Sh.			4	0
	OFF DELAWARE BAY.				1881.			
1043	38 39 00	73 11 00	130	S.	Oct. 10	5		
1046	38 33 00	73 18 00	104	S.	Oct. 10	1		
1047	38 31 00	73 21 00	156	S.	Oct. 10	1		
	OFF CHESAPEAKE BAY.				1880.			
896	37 26 00	74 19 00	56	S. Sh.	Nov. 16	7	1	1
899	37 22 00	74 29 00	57	S.	Nov. 16	11		

The type specimen of *C. depressus* which I have examined is from the Straits of Florida, and is labeled "Bache, Apr. 2, 5th cast, 54 fms., off Sombrero." This specimen gives the following measurements in millimeters and hundredths of the length of the carapax:

Sex	♂
Length of carapax	7.0
Breadth of carapax	5.2
Same in hundredths of length	74
Length of cheliped	8.0

Length of chela.. 3.3
Same in hundredths of length of carapax... 47
Height of chela.. 1.2
Same in hundredths of length of carapax... 17
Length of dactylus... 1.9

Neither Stimpson nor Milne-Edwards mentions the presence of an interantennular spine in any of the species of *Collodes*, and both of them speak of it in *Euprognatha* as specially distinguishing that genus from its near allies; but in the two species which I have examined the spine is well developed, though less prominent, and not projecting forward at all in *C. depressus*.

Euprognatha rastellifera Stimpson.

> Stimpson, Bull. Mus. Comp. Zool. Cambridge, ii, p. 123, 1870.
> A. M.-Edwards, Crust. Région Mexicaine, p. 183, pl. 33, fig. 2, 1878; Bull. Mus. Comp. Zool. Cambridge, viii, p. 7, 1880.
> Smith, Proc. Nat. Mus., iii, p. 415, 1881; Bull. Mus. Comp. Zool. Cambridge, x, p. 4, 1882.

Specimens examined.

Station No.	Locality.		Depth in fathoms.	Nature of bottom.	When collected.	No. of specimens.		With eggs.
	N. lat.	W. long.				♂	♀	
	OFF MARTHA'S VINEYARD.				1880.			
865	40 05 00	70 23 00	65	fne. S. M.	Sept. 4	100+		+
869	40 02 18	70 23 06	192	fne. S.	Sept. 4	6		....
871	40 02 54	70 23 40	115	fne. S. M.	Sept. 4	1,000+		+
872	40 05 30	70 23 52	86	S. G. Sh. sponges.	Sept. 4	50+		+
873	40 02 00	70 57 00	100	Sft. M.	Sept. 13	500+		+
874	40 00 00	70 57 00	85	Sft. M.	Sept. 13	1,000+		+
877	39 56 00	70 54 18	126	Sft. M.	Sept. 13	20+		+
878	39 55 00	70 54 15	142	M.	Sept. 13	500+		+
					1881.			
920	40 13 00	70 41 54	63	gn. M.	July 16	7		....
921	40 07 48	70 43 54	67	gn. M.	July 16	1,500+		+
922	40 03 48	70 45 54	71	gn. M. S.	July 16	250+		+
923	40 01 24	70 46 00	96	S.	July 16	37		+
925	39 55 00	70 47 00	229	S. M.	July 16	5		....
940	39 54 00	69 51 30	134	hrd. S., sponges.	Aug. 4	....		....
941	40 01 00	69 56 00	79	hrd. S. M.	Aug. 4	2,000+		+
949	40 03 00	70 31 00	100	yl. M.	Aug. 23	500+		+
950	40 07 00	70 32 00	71	S. Sh. M.	Aug. 23	150+		+
1036	39 58 00	69 30 00	94	S.	Sept. 14	1		....
1038	39 58 00	70 06 00	146	S. Sh.	Sept. 21	2		....
1046	40 00 00	70 06 00	93	S. Sh.	Sept. 21	20+		+
	OFF DELAWARE BAY.				1881.			
1043	38 39 00	73 11 00	130	S.	Oct. 10	10		....
1047	38 31 00	73 21 00	156	S.	Oct. 10	7		....
	OFF CHESAPEAKE BAY.				1880.			
896	37 26 00	74 19 00	56	S. Sh.	Nov. 16	3	1	+
899	37 22 00	74 29 00	57	S.	Nov. 16	6	4	+

I have also examined specimens taken by Alexander Agassiz on the Blake in 1880, at the following stations:

Station.	N. lat.	W. long.	Fathoms.	Specimens.
	° ′ ″	° ′ ″		
335	38 22 25	73 33 40	89	1♂
343	40 10 15	71 4 30	71	70♂ ♀
346	40 25 35	71 10 30	44	1♀

Among the vast number of specimens examined there are very few sexually immature individuals. Both sexes ordinarily attain maturity before the carapax is 6ᵐᵐ in length, and the scarcity of immature specimens in the collections may be due to their small size causing them to be overlooked in the great mass of material brought up in the trawl. The largest females seen do not exceed 10ᵐᵐ in length of carapax, and differ very little from the smallest in the form and proportions of chelipeds and ambulatory legs, though the carapax is a little broader in proportion and the spines with which it is armed are much lower, or reduced to tubercles, in the larger specimens. The males attain much greater size than the females, the carapax often exceeding 14ᵐᵐ in length, and there is a very marked and constant increase in the size of the chelipeds, particularly in the height and the thickness of the body of the chelæ, well shown in the accompanying table of measurements. In both sexes there is considerable variation in the length of the spines of the carapax, even in specimens of the same size, and there is a marked decrease in the length of the spines with the growth of the individual. In large specimens the spines upon the orbital arches, upon the gastric, cardiac, and the summits of the branchial regions, and upon the basal segment of the abdomen, are usually reduced to low, and often inconspicuous, tubercles.

The number and arrangement of the branchiæ and epipods are indicated in the following formula:

	Somite—								Total.
	VII.	VIII.	IX.	X.	XI.	XII.	XIII.	XIV.	
Epipods..................	1	0	1	0	0	0	0	0	(2)
Podobranchiæ............	0	1	1	0	0	0	0	0	2
Arthrobranchiæ.........	0	0	2	0	0	0	0	0	2
Pleurobranchiæ.........	0	0	0	2	1	1	0	0	4
									8 + (2)

The sixth and seventh somites of the abdomen of the male are anchylosed completely, as they are also in *Euprognatha rastellifera*, Col-

lodes depressus, C. robustus, and *Lispognathus furcatus,* though neither Stimpson nor Milne-Edwards mentions it, and Milne-Edwards even apparently figures them as separate in *E. rastellifera* and *C. depressus.*

Measurements in millimeters and hundredths of length of carapax.

	Station—									
	865	865	865	878	878	922	865	865	869	878
Sex	♂*	♂	♂	♂	♂	♂	♀	♀*	♀	♀
Length of carapax, including rostrum	3.2	3.1	5.6	6.8	11.3	14.4	5.8	6.0	7.2	9.5
Breadth of carapax, excluding spines	2.3	3.6	4.1	5.0	8.9	12.0	4.4	4.5	5.5	7.7
Same in hundredths of length of carapax	72	71	73	74	79	83	76	75	76	81
Length of cheliped	4.0	6.5	7.5	10.0	21.0	29.0	6.8	7.0	8.0	11.8
Length of chela	1.5	2.7	3.1	4.2	10.0	12.8	2.6	2.8	3.2	4.7
Same in hundredths of length of carapax	47	53	55	62	88	89	45	47	44	49
Height of chela	0.3	0.6	0.8	1.3	2.7	3.6	0.6	0.6	0.7	1.0
Same in hundredths of length of carapax	10	12	14	19	24	26	10	10	10	11
Length of dactylus	0.6	1.3	1.6	2.0	4.0	5.0	1.2	1.4	1.6	2.3
Length of first ambulatory leg	5.7	11.0	13.0	16.0	32.0	35.0	11.0	8.5	13.5	19.8
Length of propodus	1.5	3.1	3.5	4.4	9.5	10.0	2.8	2.0	3.3	5.3
Length of carpus	1.0	2.0	2.2	2.7	5.2	5.5	2.0	1.5	2.2	3.3
Length of fourth ambulatory leg		8.0	9.0	10.8	20.0	22.0	9.0	6.2	9.5	14.0
Length of propodus		2.3	2.5	3.2	5.8	6.0	2.5	1.6	2.5	4.0
Length of carpus		1.7	1.8	2.1	3.7	4.1	1.7	1.0	1.8	3.0

* Immature specimens; the others all adult, the females with eggs, even in the case of the smallest. The first and fourth ambulatory legs in the immature female are apparently reproduced appendages, which may, perhaps, account for the retardation in the sexual development of the individual.

The specimens in the Fish Commission collections and in the Blake collection of 1880 appear to agree much more closely with those originally described by Stimpson and those figured and described by Milne-Edwards than they do with a few Caribbean specimens which I have examined and which were labeled by Milne-Edwards as this species and returned to the Museum of Comparative Zoology. These specimens, two males and five females, are from the Blake collection of 1878–'79, station 134, off Santa Cruz, 248 fathoms, and, though fully adult, are all very much smaller than any other adult specimens examined. They are also considerably smaller than the specimens described by Stimpson or Milne-Edwards. The carapax is slightly narrower than in the northern specimens, with the tubercles of the surface larger and all the spines longer and more slender; the postorbital process is slender and spiniform instead of dentiform; there is a small conical spine, much more acute and more prominent than in the northern specimens, on the eye, at the emargination of the cornea; and the ambulatory legs are more slender and armed with small spiniform tubercles which are much more conspicuous than in the northern specimens. In the males the chelæ are proportionally larger, with the bodies stouter and more swollen; and in both sexes the chelæ and other parts of the chelipeds are armed with larger and more scattered tubercles, many of which, especially on the carpus and merus, become spiniform and conspicuous. Some of these differences are well shown in the following measurements (in millimeters and hundredths of length of carapax) of four of the specimens from off Santa Cruz:

	1.	2.	3.	4.
Sex	♂	♂	♀	♀
Length of carapax, including rostrum	5.3	5.6	5.4	6.0
Breadth of carapax, excluding spines	3.6	3.8	3.9	4.4
Same in hundredths of length	68	68	74	73
Length of cheliped	8.0	9.0	5.7	......
Length of chela	3.4	3.5	2.6	......
Same in hundredths of length of carapax	64	62	48	......
Height of chela	0.8	1.0	0.5	......
Same in hundredths of length of carapax	15	18	9	......
Length of dactylus	1.6	1.7	1.2	......

These Caribbean specimens are apparently specifically distinct, but a series of specimens from different parts of the West Indian region would perhaps show them to be a geographical or local variety.

Lispognathus furcatus A. M.-Edwards.

> *Lispognathus furcatus* A. M.-Edwards, Bull. Mus. Comp. Zool. Cambridge, vii, p. 9, 1880.
>
> ? *Lispognathus furcillatus* A. M.-Edwards, Rapport sur la Faune sous-marine dans les grandes profondeurs de la Méditerranée et de l'Océan Atlantique (Arch. Missions Sci. et Littéraires, ix), pp. 16, 39, 1882 (no description).

To this species I refer, with considerable hesitation, two specimens dredged off Martha's Vineyard: Station 951, N. lat. 39° 57', W. long. 70° 31' 30'', 225 fath., mud, Aug. 23, 1881 (male); station 1096, N. lat. 39° 53', W. long. 69° 47', 317 fath., soft green mud, Aug. 11, 1882 (female carrying eggs).

The carapax, excluding the rostral and lateral spines, is about four-fifths as broad as long in the male, and slightly broader and much thicker and more swollen in the female. The rostral horns are acicular, very slightly divergent, and slightly ascending, and in the male nearly three-tenths as long as the rest of the carapax. The three erect gastric and the postorbital spines are subequal and very slender and acute, and the postorbital spine each side is situated slightly in front of a line from the middle to the lateral gastric. The cardiac spine is considerably stouter and a little higher than the gastric spines, and either side of it on the dorsal part of the branchial region there is a much smaller erect spine, and on a line between this and the lateral gastric there is a similar spine in the female, but only a minute spine or tubercle in the male. There are two or three minute spines or tubercles on the protuberant superior lobe of the hepatic region, and about as many more back of these on the side of the branchial region, while on the inferior hepatic lobe, opposite the middle of the buccal area, there is a much larger spine directed downward, and back of this a smaller one, near the base of the cheliped. The supraorbital spine is slender and about as long as the gastric spines, and in the male the interantennular is fully as long, stouter, and directed downward and curved slightly forward. The basal segment of the antenna is irregularly armed beneath with small spines or teeth, and in the male with a slender spine at the distal end. The eyestalk is armed with a minute spine or tubercle in front, and above with a small tubercle at the emargination of the cornea. The exposed surface of the ischium and merus of the external maxillipeds is armed conspicuously with marginal and submarginal spines, of which one on the inner edge of the merus is very long.

The chelipeds in the male are stout and nearly twice as long as the carapax, including the rostral horns; the merus is a little shorter than the chela and triquetral, with all three of the angles thickly armed with very long and slender spines; the carpus is rounded externally, but armed like the merus; the chela is longer than the carapax, excluding the rostral horns, and naked and unarmed except by a few spines along the proximal part of the dorsal edge; the body is stout and swollen, and the digits slightly shorter than the body, nearly straight vertically but strongly curved laterally, very much compressed, grooved longitudinally on the sides and on the rather broad dorsal edge of the dactylus, and the prehensile edges crenately serrate and in contact throughout. In the female the chelipeds are only about once and a half as long as the carapax, including the rostral spines, much more slender than in the male, and armed with proportionally longer spines; and the chela is much shorter than the carapax, excluding the rostral horns; the body is scarcely at all swollen, and is armed with slender spines along both edges and with minute spines or tubercles on the sides, and the digits are proportionally longer and narrower than in the male.

The ambulatory legs are very long and slender, clothed to the tips of the dactyli with numerous curved setiform hairs which persistently retain mud and other foreign substances, and each is armed with a slender spine on the upper side of the distal end of the merus.

In the male the abdomen is much broader relatively to the sternum than in *Euprognatha rastellifera*, and has a low tuberculiform elevation on each somite. The first and second somites are narrow, the third broadest of all, the fourth and fifth successively a very little narrower, the fifth fully twice as broad as long, and the sixth and seventh consolidated as in *Euprognatha* and *Collodes*, together much broader than long and very broad and obtuse at the tip. The appendages of the first somite reach nearly to the tip of the abdomen, and their tips are stout and curved outward very strongly.

The eggs are numerous, nearly spherical, and approximately 0.6^{mm} in diameter in alcoholic specimens.

Measurements in millimeters.

	Station—	
	951.	1000.
Sex	♂	♀
Length of carapax, including rostral spines	12.0	12+
Length of carapax, excluding rostral spines	9.3	10.8
Breadth of carapax, including spines	7.6	9.6
Breadth of carapax, excluding spines	7.6	9.3
Breadth of front between orbits	2.0	2.1
Length of cheliped	23.0	19.0
Length of chela	10.0	8.0
Breadth of chela, excluding spines	3.1	2.1
Length of dactylus	4.6	4.0
Length of first ambulatory leg	41.0	38.0
Length of propodus	13.5	12.0
Length of dactylus	8.6	8.0
Length of second ambulatory leg	37.0	34.0
Length of propodus	10.8	9.9
Length of dactylus	7.0	6.8
Length of fourth ambulatory leg	31.0	30.0
Length of propodus	9.0	8.0
Length of dactylus	5.5	6.0

Lumbrus Verrillii Smith, Proc. National Mus., iii, p. 415, 1881.

Specimens examined.

Station No.	Locality. N. lat.	Locality. W. long.	Depth in fathoms.	Nature of bottom.	When collected.	No. of specimens. ♂	No. of specimens. ♀	With eggs.
	OFF MARTHA'S VINEYARD.				1880.			
865	40 05 00	70 28 00	65	fne. S. M.	Sept. 4	2	0	
872	40 05 30	70 23 52	86	S. G. Sh. Sponges.	Sept. 4	3	0	
940	39 54 00	69 51 30	134	hrd. S. Sponges.	1881. Aug. 4		1	0
949	40 03 00	70 30 00	100	yl. M.	Aug. 23	2		
950	40 07 00	70 32 00	71	S. Sh. M.	Aug. 23	4	2	0

Measurements in millimeters and hundredths of length of carapax.

	Station—									
	950	949	950	950	949	949	950	940	950	872
Sex	♂	♂	♂	♂	♂	♂	♀	♀	♀	♀
Length of carapax	15.7	16.7	17.3	20.5	20.9	23.0	17.	18.5	20.4	32.8
Breadth, including lateral spines	19.3	20.0	20.9	25.0	26.0	31.2	20.7	23.3	25.0	41.0
Same in hundredths of length	123	120	121	122	124	125	120	121	123	125
Breadth, including lateral spines	17.0	17.3	18.0	22.0	22.0	27.2	17.8	19.5	22.0	35.3
Length of cheliped fully extended	39.0	42.0	43.0	55.0	59.0	......	40.0	48.0	50.0	85.0
Same in hundredths of length of carapax.	248	252	249	268	282	......	235	260	245	239
Length of merus of cheliped	14.5	15.3	15.5	20.0	27.0	......	14.0	16.3	18.0	32.0
Length of propodus of cheliped	19.0	20.0	20.0	26.0	28.0	......	19.0	19.0	23.0	39.0

The specimens taken in 1881 are much smaller than the type specimens taken in 1880 ; none of the females are fully adult, and the largest males, though adult, are apparently not fully grown. The largest of the males differ very little from the females originally described, except that the chelipeds are proportionally a little larger. In the smaller specimens of both sexes there are rather fewer small tubercles upon the carapax, and the teeth of the lateral margins of the carapax and angles of the chelipeds are, perhaps, smaller and less lacineated proportionally, but the differences are very slight, and there is no approach to *L. Pourtalesii* as figured by A. Milne-Edwards. The accompanying table of measurements shows the slight variations in the proportions of the carapax and chelipeds better than description. In some specimens the chelipeds are slightly unequal, but in none conspicuously so, and when the difference was noticeable in the specimens measured the measurements of the cheliped were made from the larger one.

CANCROIDA.

Cancer borealis Stimpson.

Taken off Martha's Vineyard, in 1880, 1881, and 1882, at a great number of the stations, in 51 to 317 fathoms, and also in shallow water ; off Delaware Bay, 1881, stations 1047 and 1049, 156 and 435 fathoms ; and off Chesapeake Bay, 1880, stations 896, 897, 899, and 901, 18 to 157 fathoms. Most of the deep-water specimens taken by the Fish Commission are small, but much larger specimens, among them several from 100 to 130 millimeters in breadth of carapax, were taken in 1880, by Alexander Agassiz, on the Blake, off the Carolina coast, in 142 to 233 fathoms. The largest of these specimens were from Blake station 314 ; N. lat. 32° 24′, N. long. 78° 44′, 142 fathoms.

Cancer irroratus has not been taken in any of the deeper dredgings off Martha's Vineyard, although it is a common littoral and shallow-water species on the whole New England coast, and was taken by Alexander Agassiz at several stations, in 65 to 178 fathoms, off the Carolina coast, even occurring with *C. borealis* at station 314, just mentioned.

Geryon quinquedens Smith.

Trans. Conn. Acad., v, p. 35, pl. 9, figs. 1, 2, 1879; Proc. National Mus., iii, p. 417, 1881; Bull. Mus. Comp. Zool. Cambridge, x, p. 6, 1882.

Specimens examined.

Station No.	Locality. N. lat.	Locality. W. long.	Depth in fathoms.	Nature of bottom.	When collected.	No. of specimens. ♂	No. of specimens. ♀	With eggs.
	OFF MARTHA'S VINEYARD.							
	° ′ ″	° ′ ″			1880			
881	39 46 30	70 54 00	325	M.	Sept. 13	1		
893	39 52 20	70 53 00	372	sft. bn. M. sml. St.	Oct. 2		1	0

Specimens examined—Continued.

Station No.	Locality. N. lat.	Locality. W. long.	Depth in fathoms.	Nature of bottom.	When collected.	No. of specimens. ♂	No. of specimens. ♀	With eggs.
	OFF MARTHA'S VINEYARD Continued.				1881.			
937	39 49 25	69 49 00	616	gn. S. M.	Aug. 4	4		
945	39 58 00	71 13 00	207	gn. M. S.	Aug. 9	1	1	0
946	39 55 30	71 14 00	247	gn. M. S.	Aug. 9		1	0
947	39 53 30	71 13 30	319	S. M.	Aug. 9	2	2	0
952	39 55 00	70 28 00	396	yl. M. S.	Aug. 23		1	0
994	39 40 00	71 30 00	368	M.	Sept. 8	1		
1029	39 57 06	69 16 00	458	yl. M. S.	Sept. 14	1		
					1882.			
1124	40 01 00	68 54 00	640	fne. S. gn. M.	Aug. 26	3	2	1
1125	40 03 00	68 56 00	291	S. M.	Aug. 26			
1140	39 34 00	71 56 00	374	sft. M. P.	Sept. 8	7	2	0
1142	39 32 00	72 00 00	322	S. M. P.	Sept. 8	1	4	0
1143	39 29 00	72 01 00	452	sft. M.	Sept. 8	1		
	OFF DELAWARE BAY.				1881.			
1043	38 28 00	73 22 00	435	M.	Oct. 10	2	1	

In the Blake dredgings of 1880 the species was taken at the following
stations:

Station.	N. lat.	W. long.	Fathoms.	Specimens.
325	32 35 20	76 0 0	647	1 ♂
332	35 45 30	74 48 0	263	2 ♂
334	38 26 30	73 26 40	395	2 ♂
337	38 20 8	73 23 20	740	Fragments only.
348	39 45 40	70 55 0	732	3 ♀ with eggs.
309	40 11 40	68 22 0	304	1 ♂, 1 ♀
312	39 50 45	70 11 0	466	1 ♂

This species grows to be by far the largest brachyuran in our waters.
The largest specimen which I have seen is from the Blake collection of
1880, and was taken off Cape Hatteras. This specimen, measurements of
the carapax of which are given in the last line of the following table of
measurements, is more than six inches across the carapax and two feet
across the outstretched legs. Very large individuals differ considerably
from the specimens originally described. In all the large specimens the
teeth of the antero-lateral margin of the carapax become reduced to angular tubercles, and in some of the larger ones the fourth tooth becomes
entirely obsolete. Specimens of the same size vary much, particularly the
larger ones, in the prominence of the anterolateral teeth, so that the pro-
pertional breadth of the carapax, including the teeth or spines, varies
much more than the breadth excluding the teeth or spines, as shown
in the table of measurements. This variation is partially due to the
wearing away of the teeth, which probably takes place rapidly on
account of the softness of the exoskeleton, which is much less calcareous
than usual, the branchial regions of the carapax being so soft as to be
readily bent or indented with the finger.

Vol. VI, No. 2.   Washington, D. C.   June 18, 1883.

Measurements of the carapax in millimeters and lengths of carapax.

Station.	Sex.	Length of carapax.	Breadth, including teeth.	Breadth, excluding teeth.
		Mm.	*Mm. Length.*	*Mm. Length.*
1142	♂	11.7	15.5 = 1.32	13.9 = 1.19
947	♂	23.0	30.5 = 1.33	25.3 = 1.10
952	♂	33.0	42.0 = 1.27	36.3 = 1.10
1049	♂	35.3	44.4 = 1.26	39.0 = 1.10
947	♂	37.0	46.5 = 1.26	42.0 = 1.14
1140	♂	43.7	56.1 = 1.28	50.0 = 1.14
1140	♂	46.9	61.3 = 1.31	53.0 = 1.13
1140	♂	95.0	113.0 = 1.20	108.0 = 1.14
994	♂	97.0	114.0 = 1.18	105.0 = 1.08
937	♂	100.0	117.0 = 1.17	109.0 = 1.09
1029	♂	102.0	123.0 = 1.21	110.0 = 1.14
1140	♂	103.0	120.0 = 1.17	113.0 = 1.10
1143	♂	103.0	124.0 = 1.20	115.0 = 1.11
1140	♂	106.0	125.0 = 1.18	117.0 = 1.10
937	♂	106.0	126.0 = 1.19	115.0 = 1.08
1049	♂	114.0	132.0 -- 1.16	124.0 = 1.09
937	♂	114.0	133.0 = 1.17	125.0 = 1.09
1140	♂	114.0	129.0 = 1.13	123.0 = 1.08
937	♂	115.0	134.0 = 1.17	125.0 = 1.09
1142	♀	11.2	15.5 = 1.38	12.3 = 1.10
1049	♀	11.7	15.4 = 1.32	14.0 = 1.20
1142	♀	11.7	15.5 = 1.32	13.9 = 1.19
1142	♀	15.2	22.2 = 1.46	17.3 = 1.14
1142	♀	15.6	21.1 = 1.35	17.5 = 1.12
947	♀	37.0	48.4 = 1.31	42.0 = 1.14
1142	♀	66.0	80.0 = 1.21	73.0 = 1.10
946	♀	69.0	85.0 = 1.23	78.5 = 1.14
1140	♀	95.0	110.0 = 1.16	104.0 = 1.09
332	♂	130.0	152.5 = 1.17	144.0 = 1.11

Bathynectes longispina Stimpson.

Bathynectes longispina Stimpson, Bull. Mus. Comp. Zool. Cambridge, ii, p. 146, 1870 (young ♂); A. M.-Edwards, Crust. Région Mexicaine, p. 234, pl. 42, fig. 1, 1879 (young ♂); Smith, Proc. National Mus., iii, p. 418, 1881.
Bathynectes brevispina Stimp., loc. cit., p. 147, 1870 (large ♀); A. M.-Edwards, op. cit., p. 235, 1879 (= Stimpson).

Specimens examined.

Station No.	Locality. N. lat.	Locality. W. long.	Depth in fathoms.	Nature of bottom.	When collected.	No. of specimens. ♂	No. of specimens. ♀	With eggs.
	OFF MARTHA'S VINEYARD.				1880.			
	° ′ ″	° ′ ″						
871	40 02 54	70 23 46	115	fne. S. M.	Sept. 4		2 y.	....
872	40 05 39	70 23 52	86	S. G. Sh. sponges.	Sept. 4		2 y.	....
874	40 00 00	70 57 00	85	sft. M.	Sept. 13	1 y.		....
879	39 49 30	70 54 00	225	S. bu. M.	Sept. 13		1 y.	....

Specimens examined—Continued.

Station No.	Locality.			Depth in fathoms.	Nature of bottom.	When collected.	No. of specimens.		
	N. lat.	W. long.					♂	♀	With eggs.
	OFF MARTHA'S VINEYARD —Continued.								
	° ′ ″	° ′ ″				1881.			
940	39 54 00	69 51 30		134	hrd. S. sponges.	Aug. 4	1		....
1038	39 58 00	70 06 00		146	S. Sh.	Sept. 21	1		....
						1882.			
1097	39 54 00	69 44 00		158	Fne. S.	Aug. 11	1		....
1152	39 58 00	70 35 00		115	S.	Oct. 4	1		....
	OFF DELAWARE BAY.								
						1881.			
1043	38 39 00	73 11 00		130	S.	Oct. 10		2	0

Stimpson's *B. longispina* was based on very young males, the length of carapax in his measurement of a single specimen being equal to 14.5^{mm}, and the *B. brevispina* on a very large female in which the carapax was 49^{mm} in length. A. Milne-Edwards's specimens were evidently small, although he apparently translates the measurements given by Stimpson and does not indicate the exact size of the specimen figured. In the series of specimens which I have examined the largest are connected with the smallest by a complete series, and though none of the specimens are as large as the type of Stimpson's *brevispina*, the larger ones, both male and female, approach it closely enough in the length of the lateral spines of the carapax, etc., to make it clear that the forms described by Stimpson belong to the same species. The accompanying table of measurements will show this quite as well as any description.

In specimens shortly after being placed in alcohol, and before the colors had changed materially from those in life, the dorsum of the carapax was dull red, the color being almost wholly upon the tubercles and granules, while the ground between was grayish, though the spines and teeth of the margin were brighter red than the general surface from a slight deposit of color between the tubercles and granules. The ventral surface of the carapax, the antennulæ, antennæ, external maxillipeds, sternum, abdomen, and the proximal portions of the ambulatory legs were pale red or tinged with red. The chelipeds were specked and slightly mottled with red; the terminal third of the digits scarlet, somewhat obscured at the tips by blackish. The meral and carpal segments of the first three pairs of ambulatory legs, and the meral, carpal, and propodal segments of the posterior pair were specked and mottled with scarlet; the propodal segments of the first three pairs, except a narrow band at the distal end, and the whole of the dactyli of all four pairs were bright scarlet.

Measurements in millimeters and hundredths of length of carapax.

	Station—							
	871	879	874	1125	1038	940	1043	1043
Sex	Yng.	Yng.	♂ y.	♂	♂	♂	♀	♀
Length of carapax, including frontal teeth	8.9	9.8	13.3	21.7	23.2	29.8	26.3	35.5
Length of carapax, excluding frontal teeth	8.6	9.5	12.8	21.0	22.4	28.7	25.7	34.0
Breadth of carapax in front of lateral spines	10.3	11.5	15.8	26.2	28.0	37.0	31.4	42.7
Same in hundredths of length first given	116.	117.	119.	121.	121.	124.	119.	120.
Breadth of carapax, including lateral spines	16.2	17.0	24.2	40.8	43.8	56.0	46.0	65.0
Length of lateral spines	3.3	3.6	5.0	8.7	9.3	10.8	7.3	12.0
Length of right cheliped	14.0	15.0	21.0	35.0	39.0	53.0	44.0	60.0
Length of chela	7.7	8.0	11.3	19.0	21.3	28.5	23.6	32.0
Height of chela	2.7	2.9	3.8	6.7	7.9	10.3	8.3	12.0
Length of dactylus	3.7	3.8	6.0	9.0	10.7	14.3	12.0	16.8
Length of left cheliped	13.5	15.5	18.0	34.0	......	51.0	42.5	57.0
Length of chela	7.5	7.6	9.3	18.3	......	27.5	23.0	31.0
Height of chela	2.5	2.8	2.5	6.0	......	9.1	7.7	10.0
Length of dactylus	3.7	4.0	5.7	9.1	......	14.7	12.0	17.0
Length of third ambulatory leg	22.0	25.0	35.0	52.0	60.0	72.5	64.0	75.0
Length of fourth ambulatory leg	14.0	15.0	21.0	33.0	38.5	46.5	41.0	55.0
Length of dactylus	4.4	4.6	......	9.7	11.2	14.5	12.2	16.5
Breadth of dactylus	1.2	1.5	......	3.7	4.3	5.4	5.0	7.0

OXYSTOMATA.

Acanthocarpus Alexandri Stimpson.

Although this species occurred in considerable abundance in the dredgings off Martha's Vineyard in 1880, being taken at seven out of the fourteen stations in between 50 and 200 fathoms, it was taken but once in 1881, station 944, 128 fathoms, and was not taken at all in 1882.

In the living specimens taken in 1881 the dorsal surface of the carapax and chelipeds was pale reddish orange, deepest in color upon the elevations of the carapax and upon the bases of the carpal spines of the chelipeds; while the carapax beneath, the sternum, abdomen, and the under surfaces of the chelipeds and ambulatory legs were white, very slightly tinged with reddish.

Myropsis quinquespinosa Stimpson, Bull. Mus. Comp. Zool. Cambridge, ii, p. 157, 1870 ; A. M.-Edwards, ibid., viii, p. 21, 1880.

Station 941, N. lat. 40° 1′, W. long. 69° 56′.

A single very large male, which gives the following measurements:

Millimeters.

Length of carapax, including frontal lobes and posterior spine	37.0
Length of carapax, excluding frontal lobes and posterior spine	34.0
Breadth of carapax, including lateral tubercles	31.4
Breadth of carapax, excluding lateral tubercles	31.0
Length of cheliped	75.0
Length of merus	32.5
Length of chela	34.8
Length of dactylus	21.0
Length of first ambulatory leg	56.0
Length of posterior ambulatory leg	36.0

In life the dorsal surface of the carapax and the chelipeds and ambulatory legs are pale orange red.

Cymopolia gracilis, sp. nov.

This species, of which only one specimen has been obtained, resembles *C. cursor*, A. Milne-Edwards (Bull. Mus. Comp. Zool. Cambridge, viii, p. 29, 1880), in the great length of the second pair of ambulatory legs, but is at once distinguished by the much smoother carapax without tubercles on the posterior margin, by the broad sinuses of the superior margin of the orbit, and by the conspicuously hook-shaped tips of the first pair of abdominal appendages of the male.

Male.—The front is deeply divided by a sharp median sinus, and is slightly and obtusely bilobed either side, with the inner lobes much more prominent than the lateral. The orbit is very broad and open above. The superior margin is armed with two small teeth, separated from each other and from the inner and outer angles of the orbit by rounded sinuses, of which the inner is very broad and nearly semicircular; the middle and outer successively smaller; the outer angle is triangular and a little less prominent than the outer suborbital lobe, which is dentiform and separated from it by a shallow sinus; and the inner suborbital process (which is also the dorsal wall of the efferent branchial passage) is narrow, rounded at the tip, reaches nearly as far forward as the lobes of the front, and is separated from the outer suborbital lobe by a very broad and rounded sinus. The antero-lateral margin is unarmed, except by a small dentiform tubercle on the anterior part of the branchial region in place of the sharp tooth in *C. cursor*. The dorsal surface of the carapax is naked, minutely granulated, and armed with a very few low and obtuse tubercles. There are three faintly indicated tubercles on the middle of the gastric region; two, the largest of all, surmount a transverse ridge on the anterior part of the cardiac region; on either side, and nearly in line with these, are two smaller ones on the branchial region, above and back of the dentiform marginal tubercle already referred to; and in front of these two small ones there is a slight but scarcely tuberculiform elevation.

The eyes are large, the greatest diameter equaling nearly a third the length of the carapax, reniform, and bear upon the upper side of the stalk, near the cornea, two or three minute elevations, which are much less conspicuous than the tubercles similarly situated in *C. cursor*.

The chelipeds are slightly longer than the breadth of the carapax, and the chelæ are slender, naked, and nearly smooth, and the long, compressed, and very slender digits hooked at the tips and serrate along the prehensile edges. The first ambulatory leg is nearly twice as long as the breadth of the carapax, very slender, naked, and nearly smooth, except a very few minute granular tubercles near the base of the merus, and the dactylus is nearly as long as the propodus, subcylindrical, regularly tapered and slightly curved. The second ambulatory leg is apparently more than twice as long as the first; the merus reaches nearly to the tip of the first leg, is tapered distally, and is armed with a few minute teeth near the distal end of the posterior edge and along the anterior and dorsal surface with small granular tubercles which become obsolete distally, are much less conspicuous than in *C. cursor*, and not definitely arranged in several longitudinal lines as in that species; the carpus is about two-fifths as long as the merus, slender and unarmed; the dactylus and the distal part of the propodus are wanting. The third ambulatory leg is a little longer than the first, fully as slender, and very much like it in lack of ornamentation and in the proportions of the segments. The posterior ambulatory legs are shorter than the merus in the third, and very slender.

The abdomen is unarmed externally. The first pair of appendages reach to the second sternal somite, and the distal part of each appendage is straight to near the tip, which is curved outward and backward in a semicircular, blunt-pointed hook, and armed on the outer edge at the base of the hook with a conspicuous tooth.

Station 878, off Martha's Vineyard, 1880, N. lat. 39° 55', W. long. 70° 54' 15'', 142 fath., fine sand and mud; one specimen. The measurements in the first column of the accompanying table are from this specimen, while those in the second column are taken from one of the type specimens of *C. cursor*.

Measurements in millimeters and hundredths of length of carapax.

	C. gracilis.	C. cursor.
Sex	♂	♂
Length of carapax, including frontal teeth	5.0	6.5
Greatest breadth of carapax	7.0	8.7
Same in hundredths of length	140	134
Length of cheliped	7.3	....
Length of chela	3.1	3.6
Height of chela	0.7	0.8
Length of dactylus	1.8	1.9
Length of first ambulatory leg	13.0	11.5
Length of merus	4.5	3.8
Length of propodus	3.1	3.2
Length of dactylus	3.0	3.0
Length of second ambulatory leg	....	30.0
Length of merus	11.0	10.5
Length of carpus	4.2	4.7
Length of propodus	....	9.0
Length of dactylus	....	5.4
Length of third ambulatory leg	14.5	18.0
Length of merus	5.5	5.5
Length of propodus	3.4	5.5
Length of dactylus	3.1	3.7
Length of fourth ambulatory leg	5.0	5.5

In *C. cursor* the teeth of the superior margin of the orbit are much larger than in *gracilis* and the sinuses smaller and more triangular. The anterior-lateral margin projects in a dentiform tubercle on the hepatic region, and back of this on the anterior part of the branchial region there is an acute and prominent tooth directed somewhat forward, and a smaller but acute tooth, just back of its base.* The first pair of

---

* There is evident confusion in regard to the armament of the antero-lateral margin in Milne-Edwards's description above referred to, for he says, "Le bord latéral ne porte pas des dents, en avant du sillon post-hépatique les régions branchiales sont pourvues des quelques gros tubercules sur leur bord." I have examined four of the original specimens of *C. cursor* returned to the Museum of Comparative Zoology, and they all have the antero-lateral margin armed, as here described, but agree in all other respects with Milne-Edwards's brief description.

abdominal appendages of the male are fully as long as in *gracilis*, but the tips are slender and styliform instead of hooked.

Ethusa microphthalma Smith, Proc. National Mus., iii, p. 418, 1881.

Station 921, off Martha's Vineyard, N. lat. 40° 7' 48", W. long. 70° 43' 54", 67 fath. (1 ♂, 1 ♀); station 1047, off Delaware Bay, N. lat. 38° 31', W. long. 73° 21', 156 fath. (1 ♂). The original specimen was from station 878, off Martha's Vineyard, N. lat. 39° 55', W. long. 70° 54' 15", 142 fath.

The female from station 921 is fully adult, but does not differ essentially from the immature female from which the species was originally described; in this fully adult specimen the antero-lateral angles of the carapax, however, project farther forward, reaching a little beyond the spines of the front, and the ambulatory legs are apparently proportionally longer and have proportionally slightly longer and narrower dactyli.

The two males differ very remarkably from one another, and are possibly distinct species. The one from station 921 is only slightly larger than the immature female (from station 878) and differs very little from it in the proportions of the carapax, the form of the front, or in the eyes, external oral appendages, or ambulatory legs, except that the first and second pairs are proportionally longer, with slightly longer and narrower dactyli. The chelipeds, however, are very unequal. The left is slender throughout, and like those of the female, while the right, though very little longer than the left, has a very stout and swollen chela. The right merus is much like the left, but considerably stouter; the carpus is much stouter than the left, and considerably swollen; and the chela is more than twice as thick as the left, smooth and naked throughout, the body longer than the digits and much swollen, and the digits tapered to the tip, the prehensile edges somewhat oblique and unarmed. The male from station 1047, though of about the same size as the other, has a narrower carapax, distinctly longer than broad, but with the front absolutely broader; the ambulatory legs are considerably shorter, and with slightly broader dactyli; and the chelipeds are equal, and like the left one of the other male, except that they are very slightly shorter, and with proportionally slightly shorter chelæ.

Measurements in millimeters.

	Station—		
	921	1047	921
Sex	♂	♂	♀
Length of carapax, including median frontal spines	14.8	15.0	22.0
Greatest breadth of carapax	15.0	14.1	22.6
Breadth between antero-lateral spines	7.7	8.0	10.0
Breadth between tips of inner angles of orbital sinuses	3.5	3.7	4.4
Length of right cheliped	23.5	20.0	29.0
Length of chela	10.0	8.4	12.2
Breadth of chela	4.5	2.1	2.7
Length of dactylus	5.0	4.5	7.3
Length of left cheliped	22.0	20.0	29.0
Length of chela	9.2	8.5	12.4
Breadth of chela	2.0	2.0	2.6
Length of dactylus	5.0	4.5	7.4
Length of second ambulatory leg	50.0	45.0	68.0
Length of propodus	12.0	10.5	15.3
Length of dactylus	14.4	12.0	20.5
Length of fourth ambulatory leg	19.5	19.0	25.0
Length of propodus	4.6	4.0	5.0
Length of dactylus	1.6	1.6	2.0

In life, the carapax, the proximal part of the abdomen, the chelipeds, and first and second ambulatory legs, are pale orange, the color deepest on the chelæ and the propodi and dactyli of the ambulatory legs; the rest of the animal is grayish white and more pubescent than the more brightly colored parts.

ANOMURA.

LATREILLIDEA.

Latreillia elegans Roux.

Specimens examined.

Station No.	Locality.						Depth in fathoms.	Nature of bottom.	When collected.	No. of specimens.		With eggs.
	N. lat.			W. long.						♂	♀	
	OFF MARTHA'S VINEYARD.											
	°	′	″	°	′	″			1880.			
872	40	05	39	70	23	52	86	S. G. Sh. sponges.	Sept. 4		3	0
874	40	00	00	70	57	00	85	sft. M.	Sept. 13	fragm.		
									1881.			
940	39	54	00	69	51	30	134	hrd. S. sponges.	Aug. 4	8	10	5
1027	40	00	00	69	19	00	93	fne. S.	Sept. 14	1		
	OFF DELAWARE BAY.								1881.			
1043	38	39	..	73	11	..	130	S.	Oct. 10		1	0

HOMOLIDEA.

Homola barbata White ex Fabricius.

Specimens examined.

Station No.	Locality. N. lat.	Locality. W. long.	Depth in fathoms.	Nature of bottom.	When collected.	No. of specimens. ♂	No. of specimens. ♀	With eggs.
	OFF MARTHA'S VINEYARD.							
	° ′ ″	° ′ ″			1880.			
872	40 05 39	70 23 52	86	S. G. Sh. Sponges ..	Sept. 4	2		0
					1881.			
940	39 54 00	69 51 30	134	hrd. S. sponges......	Aug. 4	3	1	1
949	40 03 00	70 31 00	100	yl. M	Aug. 23	1y.	1	1
	OFF CHESAPEAKE BAY.							
896	37 26 00	74 19 00	56	Sh. S.............	1880. Nov. 16	1		....
899	37 22 00	74 29 00	57	S	....do ...		1	0
	OFF DELAWARE BAY.							
1043	38 39 00	73 11 00	130	S	1881. Oct. 10		1	0
1046	38 33 00	73 18 00	104	S	....do ...	1	2	1

This species is also reported from the Straits of Florida and off Barbados, by A. Milne-Edwards (Bull. Mus. Comp. Zool. Cambridge, viii., p. 33, 1880).

Four specimens give the following measurements in millimeters:

	Station— 1046	Station— 1046	Station— 940	Station— 940
Sex	♀	♂	♂	♀
Length of carapax including frontal spines	20.3	22.0	24.5	26.0
Length of carapax excluding frontal spines	19.6	21.2	23.4	25.0
Breadth of carapax including spines	17.0	17.5	19.0	22.0
Greatest breadth anteriorly excluding spines...................	15.3	16.2	17.8	18.7
Greatest breadth posteriorly excluding spines...................	15.2	16.2	17.0	18.7
Length of cheliped	33.0	40.0	51.0	43.0
Length of chela	14.0	16.0	21.0	17.6
Height of chela	5.0	6.0	6.5	6.6
Length of dactylus	7.0	7.4	9.0	8.2
Length of third ambulatory leg	45.0	43.0	58.0	57.0
Length of propodus	11.7	11.8	15.5	14.7
Length of dactylus	8.8	8.0	11.2	11.0
Length of fourth ambulatory leg	28.0	30.0	34.0	35.0
Length of propodus	7.0	7.5	8.5	8.0
Length of dactylus	3.0	3.2	3.8	3.5

RANINIDEA.

Lyreidus Bairdii Smith, Proc. National Mus., iii, p. 420, 1881.

No specimens of this species have been taken since 1880.

PORCELLANIDEA.

Porcellana Sigsbeiana A. M.-Edwards, Bull. Mus. Comp. Zool. Cambridge, viii, p. 35, 1880.

Station 940, off Martha's Vineyard, N. lat. 39° 54′, W. long. 69° 51′ 30″, 134 fathoms.

A single male, which, as the following measurements show, is much larger than the specimens described by Milne-Edwards:

Millimeters.

Length of carapax	13.0
Breadth of carapax	11.6
Length of right cheliped	25.0
Length of carpus	6.6
Length of chela	13.0
Breadth of chela	4.8
Length of dactylus	5.0
Length of left cheliped	26.0
Length of carpus	6.5
Length of chela	14.5
Breadth of chela	5.7
Length of dactylus	4.5

LITHODIDEA.

Lithodes maia Leach.

A fine specimen of this northern species was taken at station 1125, off Martha's Vineyard, N. lat. 40° 3′, W. long. 68° 56′, 291 fath., sand and mud. It gives the following measurements in millimeters:

Sex	♂
Length of carapax, including rostrum and posterior spines	83
Length of carapax, excluding rostrum and posterior spines	55
Breadth of carapax between tips of hepatic spines	47.3
Breadth of carapax between tips of branchial spines	76.4
Greatest breadth of carapax, excluding spines	53.5
Length of rostrum	26.5
Length of right cheliped	86
Length of right chela	33
Breadth of right chela	13.7
Length of dactylus of right chela	18.6
Length of left cheliped	68
Length of left chela	31
Breadth of left chela	8.8
Length of dactylus of left chela	19
Length of first ambulatory leg	150
Length of second ambulatory leg	155
Length of third ambulatory leg	153
Greatest expanse of ambulatory legs	325

Lithodes Agassizii Smith, Bull. Mus. Comp. Zool. Cambridge, x, p. 8, pl. 1, 1882.

Two very small, immature specimens of this interesting species were taken off Martha's Vineyard in 1881, station 1028, N. lat. 39° 57′, W. long. 69° 17′, 410 fath., yellow mud; and station 1029, N. lat. 39° 57′ 6″, W. long. 69° 16′, 458 fath., yellow mud. Another immature specimen and two adult females were taken by Alexander Agassiz on the Blake, in 1880; the immature specimen at station 305, N. lat. 41° 33′

15″, W. long. 65° 51′ 25″, 810 fathoms ; the two females off the Carolina coast, stations 326 and 329, 464 and 603 fath.

The species is allied to *L. maia* and *L. antarctica* in having no scale and only a single spine at the base of the antenna, and in the general form and armament of the carapax and appendages, but differs from them both conspicuously in the rostrum, which is rather short and trispinous, with the lateral spines nearly as long as the rostral spine itself. The spines upon the carapax and appendages are more numerous and much more acute than in *L. maia*, and the marginal spines of the carapax are not very much larger than the dorsal. The two adults differ remarkably from each other, and from the immature specimens, in the number and length of the spines upon the carapax and legs, the spines being fewer and very much longer and more slender in the small specimens than in the adults, and more slender and more numerous in the smaller than in the larger of the two adult specimens.

Four of the five specimens seen give the following measurements in millimeters :

	Station—			
	1029.	305.	329.	326.
Sex	Young.	Young.	♀	♀
Length of carapax, including rostrum and posterior spines	17.5	25. +	115	139
Length of carapax, excluding rostrum and posterior spines	9.1	12.6	90	123
Breadth of carapax between tips of hepatic spines	13.5	18. +	57	64
Breadth of carapax between tips of branchial spines	13.0	18. +	87	117
Greatest breadth of carapax, excluding spines	6.6	9.0	77	110
Length of rostrum	7.3	9. +	17	8
Length of spines at base of rostrum	7.4	11.5	16	7
Length of anterior gastric spines	7.0	10.5	12	5
Length of anterior cardiac spines	6.3	8.0	10	5

PAGURIDEA.

Eupagurus pubescens Brandt ex Kröyer.

This species appears to be restricted to a very narrow region south of Cape Cod. It has not been taken in over 65 fathoms off Martha's Vineyard, though common in much deeper water north of Cape Cod. None of the specimens seen are large, and all the carcinoecia are composed of *Epizoanthus Americanus* or entirely overgrown with it.

Specimens examined.

Station No.	Locality.		Depth in fathoms.	Nature of bottom.	When collected.	No. of specimens.
	N. lat.	W. long.				
	OFF MARTHA'S VINEYARD.				1861.	
	° ′ ″	° ′ ″				
918	40 20 24	70 41 30	46	gn. M.	July 16	4s.
919	40 16 18	70 41 18	53	gn. M.	July 16	2s.
921	40 07 48	70 43 54	67	gn. M.	July 16	2s.
985	41 00 00	70 49 00	26	S.	Sept. 7	20+
987	40 54 00	70 48 30	28	S.	Sept. 7	11
989	40 49 00	70 47 00	30	S.	Sept. 7	10+
990	40 44 00	70 47 00	34	gn. S. M.	Sept. 7	12

Eupagurus Kröyeri Stimpson.

Nearly all the specimens are small, and in carcinœcia composed of *Epizoanthus Americanus* or overgrown with it.

Specimens examined.

Station No.	Location.						Depth in fathoms.	Nature of bottom.	When collected.	No. of specimens.	
	N. lat.			W. long.							With eggs.
	°	′	″	°	′	″					
	OFF MARTHA'S VINEYARD.								1880.		
869	40	02	18	70	23	06	192	fne. S.	Sept. 4	30+	+
870	40	02	36	70	22	58	155	fne. S. M.	Sept. 4	30+	+
877	39	56	00	70	54	18	126	fne. S. M.	Sept. 13	40+	
878	39	55	00	70	54	15	142	M.	Sept. 13	50+	+
									1881.		
920	40	13	00	70	41	54	63	fn. M.	July 16	7	
923	40	01	24	70	46	00	98	S.	July 16	1	
924	39	57	30	70	46	00	164	S.	July 16	2	
939	39	53	00	69	50	30	264	gn. S. M.	Aug. 4	2	
945	39	58	00	71	13	00	207	gn. M. S.	Aug. 9	10+	
1025	39	49	00	71	25	00	216	gn. M.	Sept. 8	12	
1026	39	50	30	71	23	00	182	gu. M. S.	Sept. 8	5	
1032	39	56	00	69	22	00	208	yl. M.	Sept. 14	50+	+
1036	39	58	00	69	30	00	94	S.	Sept. 14	10+	
1038	39	58	00	70	06	00	146	S. Sh.	Sept. 21	34	+
									1882.		
1096	39	53	00	60	47	00	317	sft. gu. M.	Aug. 11	17	
1111	40	01	33	70	35	00	124	fne. S.	Aug. 22	30y.	
1124	40	01	00	68	51	00	640	fne. S. gn. M.	Aug. 26	3	
1125	40	03	00	68	56	00	294	S. M.	Aug. 26	1	
	BLAKE DREDGINGS; A. AGASSIZ.								1880.		
305	41	34	30	65	54	30	306	S. M. G.	....	6	
306	41	32	50	65	55	00	524	fue. dk. gy. M.	....	4	
311	39	59	30	70	12	00	143	S.	....	2	

Eupagurus politus Smith. (Pl. 4, fig. 4.)

Eupagurus, sp., Smith, Proc. National Mus., iii, p. 428, 1881.

Eupagurus politus, Smith, Bull. Mus. Comp. Zool. Cambridge, x, p. 12, pl. 2, fig. 5, 1882.

The carapax is not suddenly narrowed at the bases of the antennæ, where the breadth is equal to the length in front of the cervical suture, and not rostrated, the median lobe of the front being broadly rounded and not projecting as far forward as the external angles of the orbital sinuses, which are acute and each usually armed with a short spine.

The eyestalks, including the eyes, are nearly four-fifths as long as the breadth of the carapax in front, stout, and expanded at the very large black eyes, which are terminal, not oblique, compressed vertically, and broader than half the length of the stalks. The ophthalmic scales are small, narrow, and spiniform at the tips.

The peduncle of the antennula is about as long as the breadth of the carapax in front, and the ultimate segment about a third longer than the penultimate. The upper flagellum is much longer than the ultimate segment of the peduncle, while the lower is only about half as long as the upper, slender, and composed of ten to twelve segments. The peduncle of the antenna reaches slightly beyond the eye. The acicle is slender, slightly curved, and reaches to the tip of the peduncle, and inside its base there is a minute tooth, while outside there is a straight spine toothed or spined along its inner edge, acute at the tip and half as long as the acicle itself. The flagellum is nearly naked, and about three times as long as the carapax.

The exposed parts of the oral appendages are very nearly as in *E. bernhardus*.

The chelipeds are longer, much narrower, and more nearly equal in size than in *E. bernhardus*, and, as in that species, are almost entirely naked, but beset with numerous tubercles and low spines. The right cheliped is about as long as the body from the front of the carapax to the tip of the abdomen. The merus and carpus are subequal in length, while the chela is about once and a half as long as the carpus. The carpus and chela are rounded above and armed with numerous tubercles, which are smaller and more crowded on the chela than on the carpus, but the surface between the tubercles is smooth and polished. The dorsal surface of the carpus is limited along the inner edge by a sharp angle armed with a double line of tubercles, while the outer edge is rounded. The chela is very little wider than the carpus, and is narrowed from near the base to the tips of the digits, and both edges are rounded. The digits are rather slender, about half as long as the entire chela, slightly gaping, with acute and strongly incurved chitinous tips, and the prehensile edges armed with a very few obtuse tuberculiform teeth. The left chela is much more slender than the right, but reaches to or a little by the base of its dactylus. The carpus is slender, higher than broad, only slightly expanded distally, and with the narrow dorsal surface flattened, naked, nearly smooth, and margined either side with a single line of spiniform tubercles, while the rest of the surface is beset with low, squamiform, setiferous tubercles. The chela is about a third longer than the carpus, slender, about two and a half times as long as broad, and the dactylus about two-thirds the entire length. The dorsal and outer surface is tuberculose, and a low obtuse ridge extends from near the middle of the base along the propodal digit, which tapers from the base to the tip, while the dactylus is nearly or quite smooth except for a few fascicles of setæ, more slender than the propodal digit, and tapered only near the tip. The chitinous tips of the digits are slender, acute, and strongly incurved, and the prehensile edges are sharp, and armed with a closely set series of slender spines or setæ.

The ambulatory legs reach considerably beyond the right cheliped, and the second pair reach to the tips of the first pair. In both pairs the meri and propodi are approximately equal in length and longer than the carpi, while the dactyli are about once and a half as long as the propodi, slender, strongly curved, and distally strongly twisted. The two posterior pairs of thoracic legs and the abdominal appendages are very nearly as in *E. bernhardus.*

In life the general color of the exposed parts is pale orange, the tips of the chelæ and of the ambulatory legs white, the eyes black.

The eggs are very large, and few in number as compared with the ordinary species of the genus, being 1.0^{mm} to 1.1^{mm} in diameter in alcoholic specimens, while in *E. bernhardus* they are only 0.45^{mm} to 0.50^{mm} in diameter.

Measurements in millimeters.

	Station—				
	1028.	878.	947.	990.	876.
Sex	♂	♀	♀	♂	♂
Length of carapax along median line	12.5	13.6	14.2	16.0	21.6
Breadth of carapax in front	7.0	7.1	7.0	8.7	11.6
Length of eyestalks	5.0	5.2	5.3	6.5	7.7
Greatest diameter of eye	2.9	3.0	3.0	3.2	4.0
Length of right cheliped	34.0	35.0	41.0	41.0	63.0
Length of carpus	8.3	8.8	10.0	11.0	16.5
Length of chela	12.5	13.7	16.3	16.8	25.0
Breadth of chela	7.0	6.9	8.0	8.0	11.3
Length of dactylus	7.2	7.0	8.8	8.9	13.0
Length of left cheliped	28.0	30.0	35.0	36.0	54.0
Length of carpus	7.8	7.7	8.9	9.0	13.3
Length of chela	11.0	11.0	13.6	13.5	20.1
Breadth of chela	5.1	5.0	5.6	5.7	7.5
Length of dactylus	7.0	7.1	8.7	9.0	13.0
Length of first ambulatory leg	44.0	45.0	50.0	52.0	77.0
Length of propodus	8.9	9.0	10.4	10.3	16.0
Length of dactylus	13.0	14.5	16.1	16.8	24.0
Length of second ambulatory leg	46.0	47.0	52.0	55.0	81.0
Length of propodus	10.6	9.9	11.0	11.2	17.5
Length of dactylus	14.3	15.1	17.2	18.1	26.0

The females apparently never attain as large size as the males, but they do not seem to differ from them in the relative proportions of any of the cephalothoracic appendages.

The accompanying list of specimens examined shows that this is one of the most uniformly distributed and abundant species in from 50 to 400 fathoms from Cape Cod to the Carolina coast. I have already examined specimens from more than three-quarters of the whole number of dredgings made by the Fish Commission during the past three years within this region and between these depths.

Specimens examined.

Station No.	Locality.		Depth in fathoms.	Nature of bottom.	When collected.	No. of specimens.	
	N. lat.	W. long.					With eggs.
	OFF MARTHA'S VINEYARD.						
	° ′ ″	° ′ ″			1880.		
865	40 05 00	70 23 00	65	fne. S. M.	Sept. 4	5	
869	40 02 18	70 02 06	192	fne. S. M.	Sept. 4	50+	+

*Specimens examined—*Continued.

Station No.	N. lat.	W. long.	Depth in fathoms	Nature of bottom.	When collected.	No. of specimens.	With eggs.
		OFF MARTHA'S VINEYARD —Continued.					
	o ' "	o ' "			1880.		
870	40 02 36	70 22 58	155	M. fne. S.	Sept. 4	75+	+
871	40 02 51	70 23 40	115	M. fne. S.	Sept. 4	20+	
872	40 05 39	70 23 52	86	S. G. Sh. sponges.	Sept. 4	30+	+
873	40 02 00	70 57 00	100	sft. M.	Sept. 13	10	
874	40 00 60	70 57 00	85	sft. M.	Sept. 13	15+	
876	39 57 00	70 56 00	120	sft. M.	Sept. 13	20+	
877	39 56 00	70 54 18	126	sft. M.	Sept. 13	100+	+
878	39 55 00	70 54 15	142	M.	Sept. 13	50+	+
879	39 49 30	70 54 00	225	S. bu. M.	Sept. 13	15	
880	39 48 30	70 54 00	252	M.	Sept. 13	10	
893	39 52 20	70 58 00	372	{ sft. bn. M. and sml. St.	Oct. 2	4	
894	39 53 00	70 58 30	365	{ sft. bn. M. and snd. St.	Oct. 2	10	
895	39 56 30	70 59 45	238	sft. bn. M.	Oct. 2	20+	
					1881.		
918	40 20 24	70 41 30	46	gn. M.	July 16	3 y.	
919	40 16 18	70 41 18	53	gn. M.	July 16	2 a.	
921	40 07 48	70 43 54	67	gn. M.	July 16	12 a.	
922	40 03 48	70 45 54	71	gn. M. and S.	July 16	7 l.	0
923	40 01 24	70 46 00	98	S.	July 16	12	1
924	39 57 30	70 46 00	164	S.	July 16	8	
925	39 55 00	70 47 00	229	S. and M.	July 16	3	
939	39 53 00	69 50 30	264	gn. M. and S.	Aug. 4	20 l.	2
940	39 54 00	69 51 30	134	hrd. S. and sponges.	Aug. 4	28 l.	3
941	40 01 00	69 56 00	79	hrd. S. and M.	Aug. 4	18 l.	4
943	40 00 00	71 14 30	157	M. S. and Sh.	Aug. 9	1	
944	40 01 00	71 14 30	128	M. S. and Sh.	Aug. 9	13	3
945	39 58 00	71 13 00	207	gn. M. and S.	Aug. 9	16	1
946	39 55 30	71 14 00	247	gn. M. and S.	Aug. 9	10	3
947	39 53 30	71 13 30	319	S. and M.	Aug. 9	48 l.	0
949	40 03 00	70 31 00	100	yl. M.	Aug. 23	34	5
950	40 07 00	70 32 00	71	S. Sh. and M.	Aug. 23	12	1
951	39 57 00	70 31 30	225	M.	Aug. 23	6	
990	40 44 00	70 47 00	34	gn. M. and S.	Sept. 7	2	
994	39 40 00	71 30 00	368	M.	Sept. 8	6	
997	39 42 00	71 32 00	335	yl. M.	Sept. 8	35	
*998	39 43 00	71 32 00	302	gn. M.	Sept. 8	110	
999	39 45 13	71 30 00	266	gn. M.	Sept. 8	4	
1025	39 49 00	71 25 00	216	gn. M.	Sept. 8	10	3
1026	39 50 30	71 23 00	182	gn. M. and S.	Sept. 8	25	
1027	40 00 00	69 19 00	93	fne. S.	Sept. 14	4 s.	
1028	39 57 00	69 17 00	410	yl. M.	Sept. 14	3	
1029	39 57 06	69 16 00	458	yl. M. S.	Sept. 14	1	0
1032	39 56 00	69 22 00	208	yl. M.	Sept. 14	18	
1035	39 57 00	69 28 00	120	S.	Sept. 14	5	
1036	39 58 00	69 30 00	94	S.	Sept. 14	6	
1039	39 59 00	70 06 00	130	S. and Sh.	Sept. 21	17 l.	10
					1882.		
1091	40 03 00	69 44 00	65	gy. S. brk. Sh.	Aug. 11	3	
1092	39 58 00	69 42 00	202	gy. S.	Aug. 11	24	+
1093	39 56 00	69 45 00	349	bu. M. S.	Aug. 11	4	
1096	39 53 00	69 47 00	317	sft. gn. M.	Aug. 11	14	
1097	39 54 00	69 44 00	158	fne. S.	Aug. 11	39	+
1098	39 53 00	69 43 00	156	fne. S.	Aug. 11	27	+
1108	40 02 00	70 37 30	161	gy. M. fne. S.	Aug. 22	19	+
1109	40 03 00	70 38 00	89	gy. S.	Aug. 22	58	8
1110	40 02 00	70 35 00	190	gn. M. fne. S.	Aug. 22	50+	10
1111	40 01 33	70 35 00	124	fne. S.	Aug. 22	40+	+
1112	39 56 00	70 35 00	245	gn. M. S.	Aug. 22	5	
1116	39 59 00	70 44 00	144	gn. M. S.	Aug. 23	15	+
1117	40 02 00	70 45 00	89	fne. S.	Aug. 22	5	
1118	40 03 00	70 45 00	79	fne. S.	Aug. 22	12	2
1119	40 08 00	68 45 00	97	S. brk. Sh.	Aug. 22	15	
1121	40 04 00	68 49 00	234	fne. S. St.	Aug. 26	10	+
1124	40 01 00	68 54 00	640	fne. S. gn. M.	Aug. 26	5	
1137	39 40 00	71 52 00	173	fne. S. P.	Sept. 8	1	
1138	39 39 00	71 54 00	108	fne. S. P.	Sept. 8	9	
1142	39 32 00	72 00 00	322	M. with S. and P.	Sept. 8	19	0

Specimens examined—Continued.

Station No.	Locality. N. lat.	Locality. W. long.	Depth in fathoms.	Nature of bottom.	When collected.	No. of specimens.	With eggs.
	OFF MARTHA'S VINEYARD —Continued.				**1881.**		
1152	39 58 00	70 35 00	115	S.	Oct. 4	8	
1154	39 55 31	70 39 00	193	S. and M.	Oct. 4	200+	
	OFF DELAWARE BAY.				**1881.**		
1043	38 39 00	73 11 00	130	S.	Oct. 10	2	
1045	38 35 00	73 13 00	312	gy. M.	Oct. 10	8	
1046	38 33 00	73 18 00	104	S.	Oct. 10	3	1
1047	38 31 00	73 21 00	156	S.	Oct. 10	9	
1049	38 28 00	73 22 00	435	M.	Oct. 10	1	
	OFF CHESAPEAKE BAY.						
896	37 26 00	74 19 00	56	S. Sh.	Nov. 16	3	
897	37 25 00	74 18 00	157	S. M.	Nov. 16	33	+
898	37 24 00	74 17 00	300	M.	Nov. 16	48	+
	BLAKE DREDGINGS; A. AGASSIZ.				**1880.**		
309	40 11 40	68 22 00	304	fne. S. M.		3	
310	39 59 16	70 18 30	260	fne. dk. gn. M.		2	
327	34 00 30	76 10 30	178	Glob. ooze.		1	
336	38 21 50	73 32 00	197	Bl. M.		5	

Catapagurus, A. M.-Edwards.

> *Catapagurus* A. M.-Edwards. Bull. Mus. Comp. Zool. Cambridge, viii, p. 46, 1880.—Smith, ibid., x, p. 14, 1882.
>
> *Hemipagurus* Smith, Ann. Mag. Nat. Hist. London, V, vii, p. 143, 1881; Proc. National Mus., iii, p. 422, 1881.

Catapagurus Sharreri, A. M.-Edwards. (Pl. 4, Fig. 5.)

> *Catapagurus Sharreri* A. M.-Edwards, Bull. Mus. Comp. Zool. Cambridge, viii, p. 46, 1880.
>
> *Hemipagurus socialis* Smith, Proc. National Mus., iii, p. 423, 1881.
>
> *Catapagurus socialis* Smith, Bull. Mus. Comp. Zool. Cambridge, x, p. 16, 1882.

I have examined one of the type specimens of Milne-Edwards's species returned to the Museum of Comparative Zoology, and find it identical with my species as indicated above. This specimen is from 200 fathoms, off Barbadoes, station 296, and gives the following measurements in millimeters:

Sex	♂
Length from front of carapax to tip of abdomen	23.0
Length of eye-stalks	2.3
Greatest diameter of eye	1.7
Length of right cheliped	19.0
Length of chela	8.0
Breadth of chela	2.6
Length of dactylus	4.0
Length of left cheliped	31.0
Length of chela	7.5
Breadth of chela	1.3
Length of dactylus	2.8
Length of first ambulatory leg, right side	22.0

Specimens examined.

Station No.	Locality. N. lat.	W. long.	Depth in fathoms.	Nature of bottom.	When collected.	No. of specimens.	With eggs.
	OFF MARTHA'S VINEYARD.				1880.		
865	40 05 00	70 23 00	65	fne. S. M.	Sept. 4	6	
870	40 02 36	70 22 58	155	fne. S. M.	Sept. 4	50+	+
871	40 02 54	70 23 40	116	fne. S. M.	Sept. 4	500+	+
872	40 05 39	70 23 52	86	S. G. Sh. Sponges.	Sept. 4	20+	+
873	40 02 00	70 57 00	100	sfr., M.	Sept. 13	1	
874	40 00 00	70 57 00	85	sfr., M.	Sept. 13	100+	+
876	39 57 00	70 56 00	120	sfr., M.	Sept. 13	50+	+
877	39 56 00	70 54 18	126	sfr., M.	Sept. 13	200+	+
878	39 55 00	70 54 15	142	M.	Sept. 13	50+	+
880	39 48 30	70 54 00	252	M.	Sept. 13	2	
					1881.		
919	40 16 18	70 41 18	53	gn. M.	July 16	2	
920	40 13 00	70 41 54	63	gn. M.	July 16	2	
921	40 07 48	70 43 54	67	gn. M.	July 16	12	+
922	40 03 48	70 45 54	71	gn. M. S.	July 16	48	+
923	40 01 24	70 46 00	98	S.	July 16	5	+
925	39 55 00	70 47 00	229	S. and M.	July 16	9	+
939	39 53 00	69 50 30	264	gn. S. M.	Aug. 4	1	
940	39 54 00	69 51 30	134	brd. S. sponges.	Aug. 4	1000+	+
941	40 01 00	69 56 00	79	hrd. S. M.	Aug. 4	200+	+
949	40 03 00	70 31 00	100	yl. M.	Aug. 23	15	
1027	40 00 00	69 19 00	93	fne. S.	Sept. 14	37	
1035	39 57 00	69 28 00	120	S.	Sept. 14	200+	
1036	39 58 00	69 30 00	94	S.	Sept. 14	50+	
1038	39 58 00	70 06 00	146	S. and Sh.	Sept. 21	00+	+
					1882.		
1092	39 58 00	69 42 00	202	gy. S.	Aug. 11	2	
1097	39 54 00	69 44 00	158	fne. S.	Aug. 11	3	
1111	40 01 33	70 35 00	124	fne. S.	Aug. 32	13	
1119	40 08 00	68 45 00	97	S. brk. Sh.	Aug. 26	7	
1151	39 58 30	70 37 00	126	S.	Oct. 4	10	
1152	39 58 00	70 35 00	115	S.	Oct. 4	12	
	OFF DELAWARE BAY.				1881.		
1043	38 39 00	73 11 00	130	S.	Oct. 10	3	
1047	38 31 00	73 21 00	156	S.	Oct. 10	10	
	OFF CHESAPEAKE BAY.				1880.		
899	37 22 00	74 20 00	57	S.	Nov. 16	1	
	BLAKE DREDGINGS; A. AGASSIZ.				1880.		
311	39 59 30	70 12 00	143	gy. S.		6	
313	32 31 50	78 45 00	75	fne. gy. S.		2	
314	32 24 00	78 44 00	142	fne. gy. S.		1000+	+
315	32 18 20	78 43 00	225	fne. gy. S.		4	
316	32 07 00	78 37 30	229	P.		1	
327	34 00 30	76 10 30	178	Glob. ooze.		8	
344	40 01 00	70 58 00	129	fne. S. M.		40+	+
345	40 10 15	71 04 30	71	gn. M. brk. Sh. S.		5	

Vol. VI, No. 3.   Washington, D. C.   June 18, 1883.

Catapagurus gracilis Smith.

Hemipagurus gracilis Smith, Proc. National Mus., iii, p. 426, 1881.
Catapagurus gracilis Smith, Bull. Mus. Comp. Zool. Cambridge, x, p. 19, 1882.

Specimens examined.

Station No.	Locality. N. lat.	Locality. W. long.	Depth in fathoms.	Nature of bottom.	When collected.	No. of specimens.	With eggs.
	OFF MARTHA'S VINEYARD.				1880.		
865	40 05 00	70 23 00	65	fne S. M.	Sept. 4	1	
870	40 02 36	70 22 58	155	fne. S. M.	Sept. 4	4	
871	40 02 54	70 23 40	115	fne. S. M.	Sept. 4	30+	+
874	40 00 00	70 57 00	85	sfr. M.	Sept. 13	30+	
877	39 56 00	70 54 18	126	sfr. M.	Sept. 13	3	
878	39 55 00	70 54 15	142	M.	Sept. 13	10	
919	40 16 18	70 41 18	53	gn. M.	July 16	1	
920	40 13 00	70 41 54	63	gn. M.	July 16	4	
921	40 07 48	70 43 54	67	gn. M.	July 16	24	+
940	39 54 00	69 51 30	134	brd. S. and sponges.	Aug. 4	2	
949	40 03 00	70 31 00	100	yl. M.	Aug. 23	12	
1038	39 58 00	70 06 00	146	S. Sh.	Sept. 21	1	
	OFF CHESAPEAKE BAY.				1880.		
896	37 26 00	74 19 00	56	S. Sh.	Nov. 16	1	
899	37 22 00	74 29 00	57	S.	Nov. 16	1	
	BLAKE DREDGINGS; A. AGASSIZ.						
344	40 01 00	70 58 00	129	fne. S. M.		1	
345	40 10 15	71 04 30	71	gn. M. brk. Sh. S.		3	

Parapagurus pilosimanus Smith, Trans. Conn. Acad. New Haven, v, p. 51, 1879;

Proc. National Mus. Washington, iii, p. 423, 1881; Bull. Mus. Comp. Zool.
Cambridge, x, p. 20, pl. 2, fig. 4–4d, 1882.

(Pl. 5, Figs. 3–5; Pl. 6, Figs. 1–4a.)

Specimens examined.

Station No.	Locality. N. lat.	Locality. W. long.	Depth of fathoms.	Nature of bottom.	When collected.	No. of specimens. ♂	No. of specimens. ♀	With eggs.
	GLOUCESTER FISHERIES.				1878.			
	Off Nova Scotia, 42° 41′ N., 63° 6′ W.		250			1	1	0
	OFF MARTHA'S VINEYARD.				1880.			
880	39 48 30	70 54 00	252	M.	Sept. 13	2	1 y. 1	1
893	39 52 20	70 58 00	372	sft. bn. M. and sml. St.	Oct. 2	2	1 y.	

Specimens examined—Continued.

Station No.	Locality.					Depth in fathoms.	Nature of bottom.	Where collected.	No. of specimens.	
	N. lat.			W. long.						With eggs.
	°	′	″	°	′	″					
	OFF MARTHA'S VINEYARD— Continued.								1880.		
894	39	53	00	70	58	30	365	(sft. bn. M. and sml. St.	Oct. 2	1 3	0
									1881.		
938	39	51	00	69	49	15	317	gn. S. M.	Aug. 4	3 1	0
947	39	53	30	71	13	30	319	S. M.	Aug. 9	148 245	191
994	39	40	00	71	30	00	368	M.	Sept. 8	1	
997	39	42	00	71	32	00	3.5	yl. M.	Sept. 8	1	
998	39	43	00	71	32	00	302	gn. M.	Sept. 8	1	0
1029	39	57	00	69	16	00	458	yl. M. S.	Sept. 14	1 y.	
									1882.		
1124	40	01	00	68	54	00	640	fne. S. gn. M.	Aug. 26	10	
1140	39	34	00	71	56	00	374	fne. S. slt. M. P.	Sept. 8	1	
	OFF CHESAPEAKE BAY.										
									1880.		
898	37	24	00	74	17	00	300	M.	Nov. 16	4	
	BLAKE DREDGINGS; A. AGASSIZ.										
									1880.		
306	41	32	50	65	55	00	524	fne. dk. gr. M.		1 y.	
309	40	11	40	68	22	00	304	dk. gy. S. M.		4	
322	33	10	00	76	32	15	362	Glob. S.		2	

The large number of specimens which have been obtained since this species was first described enables me to supplement to a considerable extent the original description, drawn from a single specimen from which the oral appendages were not removed.

The labrum, metastome, mandibles, and the first maxilla are essentially as in *Eupagurus bernhardus*. The lobes of the protognath of the second maxilla are very nearly as in *Eupagurus bernhardus*; the endognath is a little longer than in that species, reaching nearly as far forward as the distal lobe of the protopod; the scaphognath is very different from that of *Eupagurus bernhardus*, the anterior part being very much larger and narrowed to a triangular tip reaching much beyond the middle of the endognath, while the posterior part is elongated, somewhat ovate in outline, about two-thirds as long as the anterior, and very little more than half as broad as long. The lobes of the protopod and the endopod of the first maxilliped are nearly as in *Eupagurus bernhardus* except that the endopod is united with the exopod for a considerable distance from the base; the endopod itself, however, is very different, being a simple, unsegmented lamella, shorter than the endopod, broad and truncated at the extremity and setigerous along the outer and terminal edges. Just back of the base of the exopod the edge of the protopod is setigerous and projects laterally in a slight prominence apparently representing the epipod. The second and third (external) maxillipeds are essentially as in *Eupagurus bernhardus*.

The branchiæ are the same in number and arranged in the same way as in *Eupagurus bernhardus*, as indicated in the following formula :

	Somite—								
	VII.	VIII.	IX.	X.	XI.	XII.	XIII.	XIV.	Total.
Epipods	0	0	0	0	0	0	0	0	0
Podobranchiæ..............	0	0	0	0	0	0	0	0	0
Arthrobranchiæ............	0	0	2	2	2	2	2	0	10
Pleurobranchiæ............	0	0	0	0	0	0	1	0	1
									11

But, as stated in the original description, they are trichobranchiæ, not phyllobranchiæ as in ordinary Paguroids. In the original specimen, and in all those not preserved with special care, the branchiæ are flaccid and the papillæ of which they are composed are collapsed, apparently cylindrical throughout, and without definite arrangement along the stem of the branchia; but in specimens carefully preserved in strong alcohol the papillæ in the thicker parts of the branchiæ are seen to be slightly flattened toward their bases in the direction of the axes of the branchiæ, and to have a definite arrangement in four longitudinal series, showing, in a transverse section of the branchia, two papillæ either side of the central axis in place of the thin lamella attached by one edge to either side of the lamelliform central stem of the phyllobranchia of ordinary Paguroids. Toward the tips of the branchiæ the papillæ become truly cylindrical as in *Homarus* or *Astacus*, and in some of the smaller branchiæ, as in the arthrobranchiæ of the external maxillipeds, the papillæ upon one side of the branchia are very small or rudimentary; but in all cases the ultimate divisions of the branchiæ are apparently strictly tricho-branchial in structure, the blood vessels on either side of each papilla giving off capillary branches in opposite directions to the surface of the papilla. The structure is essentially as in *Astacus*, and the difference is not apparent without close examination. From ordinary Paguroids, like *Eupagurus bernhardus*, however, it is widely different, but this difference is partially bridged by the structure of the branchiæ in *Sympagurus pictus* about to be described, although there the branchiæ are essentially phyllobranchiæ.

In the chelipeds the merus, carpus, and chela are very densely clothed, except at the tips of the digits, a space on the under side and at the base of the chela, and the inner side of the merus, with a very fine and soft pubescence usually loaded with fine mud when the specimens are first taken.

Individuals differ considerably in the form and proportions of the chelipeds. In one large male, measurements of which are given in the last column in the accompanying table of measurements, the right cheliped is only very slightly longer and scarcely stouter than the left, and the chela differs from that of the left only slightly in form. The defective development of the right cheliped in this specimen probably resulted from the loss and reproduction of the limb, but in other specimens there are considerable differences in the form of the right chela which are apparently not the result of loss and reproduction, though it may be possible that all the cases of considerable variation in the form of the chelæ are due to this cause. The right chela is, in both sexes, usually very broad, half or more than half as broad as long, but in some specimens, as shown in the second column of the table of measurements, it is much narrower, only about three-eighths as broad as long.

The appendages of the second abdominal somite of the male are frequently very distinctly unequal in size, the right being longer than the left, but in many specimens they are exactly alike. The appendages of the first somite are apparently perfectly symmetrical in all the specimens examined.

The females appear to be a little smaller than the males, but apparently do not differ in the form or proportions of any of the cephalothoracic appendages. There are four well-developed biramus appendages on the left side of the abdomen as in the species of *Eupagurus*, and the third, fourth, and fifth somites are each furnished with a diffuse dorsal tuft of long hairs. The eggs are nearly spherical and larger than in *Eupagurus bernhardus*, being nearly a millimeter in diameter in alcoholic specimens.

In life the general color of the naked and exposed parts is pale, dull orange, darker at the tips of the ambulatory legs, without any of the conspicuous red markings characteristic of *Sympagurus pictus*.

All of the carcinœcia seen are formed by colonies of *Epizoanthus paguriphilus* Verrill, which at first invest spiral shells which are finally absorbed by the basal cœnenchyma of the growing polyps. In some of the very small specimens the investing walls of the polyp are so thin that the form and markings of the inclosed shell are distinctly visible through them, but in all the larger specimens the shell is completely absorbed.

Measurements.

	Station—				
	947.	947.	894.	947.	647.
Sex	♂	♂	♀	♀	♂
Length front to tip of telson	62.0	65.0	38.0	60.0	60.0
Length of carapax along dorsal line	23.3	23.0	15.0	18.8	22.5
Breadth of carapax at bases of antennæ	13.0	13.3	9.0	11.3	13.0
Length of eyestalks	6.4	6.7	4.7	6.0	6.3
Greatest diameter of eye	1.2	1.3	1.0	1.1	1.3
Length of right cheliped	68.0	66.0	41.0	48.0	50.0
Length of carpus	20.0	18.0	11.0	12.5	13.0
Length of chela	29.0	27.5	17.0	20.0	19.0
Breadth of chela	15.0	10.5	10.8	12.0	7.2
Length of dactylus	15.5	14.3	10.3	11.0	10.5
Length of left cheliped	51.0	52.0	30.0	35.0	49.0
Length of carpus	12.5	13.0	7.0	9.3	13.0
Length of chela	16.0	17.2	9.9	11.3	15.7
Breadth of chela	7.0	7.6	4.5	5.5	7.0
Length of dactylus	9.3	9.8	6.1	6.8	9.1
Length of first ambulatory leg, right side	98.0	100.0	58.0	63.0	96.0
Length of propodus	23.0	23.0	13.0	14.2	23.2
Length of dactylus	31.0	33.0	17.5	18.3	31.5

Sympagurus, gen. nov.

The single species of the genus here proposed is readily distinguished from *Parapagurus* by the shortness of the peduncles of the antennulæ and the well developed eyes, in which respects it agrees essentially with *Eupagurus*. It differs essentially from *Parapagurus* in having phyllobranchiæ, which are the same in number and arranged in the same way as in *Parapagurus* and *Eupagurus*, but differ much from the branchiæ of *Eupagurus* and the ordinary Paguroids in having the lamellæ long, narrow, attached by one end to the narrow stem of the branchia and arranged in two loosely packed longitudinal series either side of the axis of the branchia. At the extremity of the branchiæ, however, the lamellæ become very narrow, and at the extreme tips apparently papilliform as at the tips of the branchiæ of *Parapagurus*. The oral, thoracic. and abdominal appendages are essentially as in *Parapagurus*, the sexual appendages of the first and second somites of the abdomen of the male are, however, much smaller and less perfectly developed.

Sympagurus pictus, sp. nov. (Pl. 5, Figs. 2, 2a; Pl. 6, Figs. 5–8.)

The carapax is divided by a deep, cervical suture, which is arcuate as in *Parapagurus pilosimanus*, but is narrowed anteriorly much more than in that species, the breadth at the bases of the antennæ scarcely equaling the length in front of the cervical suture. The anterior margin projects in a prominent triangular rostrum with a distinct longitudinal carina, and either side is considerably oblique, with only a slight prominence between the base of the eyestalk and the peduncle of the antenna.

The eyestalks, including the eyes, are about two-fifths as long as the carapax along the dorsal line, stout, and expanded at the very large black eyes, which are terminal, not oblique, compressed vertically. and from two-fifths to nearly a half as broad as the length of the stalks. The ophthalmic scales are small, spiniform, and acute as in *Parapagurus pilosimanus*.

The peduncle of the antennula is a little longer than the breadth of the carapax in front, the second segment reaches to the tip of the eye, and the ultimate segment is about half the entire length. The upper flagellum is about as long as the ultimate segment of the peduncle, while the lower is only about half as long, slender, and composed of seven or eight segments. The peduncle of the antenna reaches slightly by the eye and the ultimate segment is nearly twice as long as the penultimate. The acicle is slender, sparsely setigerous, and reaches to the tip of the peduncle, and outside its base there is a dentiform process, but no tooth or spine inside. The flagellum is nearly naked and about four times as long as the carapax.

The oral appendages are all nearly as in *Parapagurus pilosimanus*, except that, in the second maxilla, the endognath is broader at the base, the anterior lobe of the scaphognath is shorter and broader, though still triangular at the tip, and the posterior lobe is shorter, broader, and approximately triangular; while, in the first maxilliped, the endopod and exopod are a little shorter and the latter rounded at the extremity.

The chelipeds are densely pubescent, as in *Parapagurus pilosimanus*, and resemble those of that species closely until the pubescence is removed, when they are seen to be different in form and armament. The right cheliped in fully grown specimens is about three times as long as the carapax along the dorsal line. The carpus is slightly longer than the merus, obscurely angulated along the inner dorsal edge, and the dorsal surface covered with small tubercles which are acute and almost spiniform along the inner edge. The chela is at least once and two-thirds as long as the carpus, much less than half as broad as long, compressed vertically, convex, and only slightly tuberculous above and below, but armed along the edges with sharp tubercles, which are most conspicuous along the inner edge and particularly on the dactylus, where they become spiniform. The digits are longitudinal, not turned to the right as in *Parapagurus pilosimanus*, about as long as the body of the chela, regularly tapered toward the strongly hooked tips, and the prehensile edges armed with irregular, low, and obtuse tubercles. The left cheliped is about two-thirds as long as the right, very slender, and clothed with pubescence like the right. The carpus is scarcely longer or stouter than the merus, and angulated and armed with a few sharp tubercles along the inner dorsal edge. The chela is about once and two-thirds as long as the carpus, scarcely stouter, rounded and unarmed, with the digits much longer than the body, slender, slightly curved downward at the tips, not gaping, and the prehensile edges sharp and armed with a closely set series of minute spines.

The ambulatory legs reach to or a little by the right cheliped, are smooth and nearly naked, except near the tips, and unarmed, except a small dentiform tooth at the distal end of the dorsal edge of the carpus. The dactyli are longer than the propodi, slender, laterally compressed, strongly curved toward the acute tips, and setigerous along the dorsal edge and on the inner side. The fourth and fifth pairs of legs and the sterna of all the thoracic somites are as in *Parapagurus pilosimanus*.

The appendages of the first and second abdominal somites of the male arise in the same way as in *Parapagurus pilosimanus*. The appendages of the first somite are like those of *Parapagurus pilosimanus* in form, but are very much smaller, being scarcely 3½ millimeters in length in the largest specimen examined, and project only a little way below the coxæ of the posterior thoracic legs. The appendages of the second somite are very unequally developed; the right is nearly as in *Parapagurus pilosimanus* in form, but is much smaller, being only 7 millimeters long in the largest male examined, and the terminal lamelliform segment is a little broader in proportion, being about a fourth longer than the basal portion and a fourth as broad as long, and is apparently less deeply grooved; while the left is very much smaller, only 4.8 millimeters long in the specimen just referred to, and the terminal lamella smaller even than the basal portion, very narrow, and scarcely at all grooved. The appendages of the left side of the third, fourth, and fifth somites of the abdomen of the male, the four ovigerous appendages of the left side of the abdomen of the female, and the uropods in both sexes, are as in *Parapagurus pilosimanus* and *Eupagurus bernhardus*. The telson is about as broad as long, but bilaterally unsymmetrical, the left side being longer than the right, and the posterior margin oblique, with a slight anal emargination a little to the right of the center.

The carcinœcium of the specimen from station 895 is formed by *Epizoanthus Americanus* Verrill, but the carcinœcia of all the other specimens examined are formed by the base of a single polyp of *Urticina consors* Verrill (Amer. Jour. Sci., III, xxiii, p. 225, 1882).

Measurements.

	Station—		
	939.	924.	1114.
Sex	♂	♂	♀
Length from front to tip of telson	27. 0	50. 0	54. 0
Length of carapax along dorsal line	10. 0	18. 0	20. 0
Breadth of carapax at bases of antennæ	5. 5	9. 8	11. 0
Length of eye-stalks	4. 0	7 0	8. 0
Greatest diameter of eye	1. 9	2. 8	3. 1
Length of right cheliped	23. 6	54. 0	60. 0
Length of carpus	6. 0	13. 0	13. 5
Length of chela	10. 0	22. 0	24. 0
Breadth of chela	4. 6	10. 0	10. 5
Length of dactylus	5. 1	11. 0	12. 0
Length of left cheliped	18. 5	35. 0	40. 0
Length of carpus	4. 7	8. 8	10. 0
Length of chela	7. 0	12. 5	14. 5
Breadth of chela	2. 6	4. 5	5. 0
Length of dactylus	5. 0	9. 0	10. 0
Length of first ambulatory leg, right side	32 0	60. 0	69. 0
Length of propodus	7. 5	13. 7	16. 0
Length of dactylus	9. 7	10. 8	19. 2

In the large male from station 924, the appendage of the right side of the second somite of the abdomen is 7^{mm} long, and its terminal lamella 4^{mm} long and 1^{mm} broad; while the appendage of the left side is 4.8^{mm} long, and its terminal lamella only 2.3^{mm} long and 0.5^{mm} broad.

In life the front part of the carapax is orange red bordered with white along the margin. The eye-stalks and the peduncles of the attennulæ and antennæ are white, except the undersides of the eye-stalks, which are vermilion. The flagella of the antennulæ and antennæ are pale orange. A large spot of vermilion covers nearly the whole of the outer surface and extends over upon the inferior edge of the meri of the ambulatory legs, and the inferior edges of the carpi and propodi and the tips of the dactyli are marked with the same color, while the rest of the surface is white. The posterior part of the carapax and the abdomen are translucent whitish specked above with orange red, and the telson and uropods are similarly but more thickly specked with the same color. The eyes are black.

Specimens examined.

Station No.	Locality.		Depth in fathoms.	Nature of bottom.	When collected.	No. of specimens.			
	N. lat.	W. long.				♂	♀	With eggs.	Dry or alc.
	OFF MARTHA'S VINEYARD.								
	° ′ ″	° ′ ″			1880.				
805	39 56 30	70 59 45	238	sft. M.	Oct. 2	1 s.			Alc.
					1881.				
924	39 57 30	70 46 00	164	S.	July 16	2 l., s.			Alc.
939	39 53 00	69 50 30	264	gn. M. S.	Aug. 4	2 s.	1 s.	0	Alc.
					1882.				
1114	39 58 00	70 28 00	171	gn. M.	Aug. 22	1 l.	1 l.	0	Alc.

GALATHEIDEA.

Munida Caribæa? Smith. (Pl. 3, Fig. 11.)

Munida Caribæa? Smith, Proc. National Mus., iii, p. 423, 1881.
Munida, sp. indet. Smith, Bull. Mus. Comp. Zool. Cambridge, x, p. 22, pl. 10, fig. 1, 1882.
? *Munida Caribæa* Stimpson, Ann. Lyceum Nat. Hist. New York, vii, p. 244 (116), 1860.—A. M.-Edwards, Mus. Comp. Zool. Cambridge, viii, p. 49, 1880 (*Caribæa*).

In my preliminary notice of two years ago I referred this species doubtfully, as indicated above, to Stimpson's species described from a single very small specimen which is no longer extant. Almost simultaneously Milne-Edwards published ten new species of the genus from the Blake dredgings in the Caribbean region, and referred specimens of still another to Stimpson's *Caribæa*, but without describing them at all. It seems best to restrict Stimpson's name to the species called *Caribæa* by Milne-Edwards, whatever that may be, but it is quite impossible to determine from Milne-Edwards's descriptions alone whether the species which I have called *Caribæa* belongs to either of the eleven species enumerated by him and, until it is possible to settle this point satisfactorily, the species may be conveniently designated *Munida Caribæa?* Smith, as above.

The species attains greater size than any of the specimens taken in 1880, measurements of some of the largest of which were given in my preliminary notice of two years ago. The specimens from the same station are usually approximately alike in size, those from one station being nearly all small, while those from another, even near by and on the same day, are nearly all large. The largest specimens are from station 1043, off Delaware Bay, and six of these give the following measurements in millimeters:

	1.	2.	3.	4.	5.	6.
Sex	♂	♂	♂	♂	♂	♂
Length	52.0	51.0	47.0	57.0	57.0	62.0
Length of carapax including rostrum	26.3	25.5	24.9	28.3	29.3	30.0
Length of rostrum	9.4	9.5	8.4	9.2	10.1	9.5
Breadth of carapax in front of cervical suture	12.2	11.7	11.8	13.4	13.4	14.4
Greatest breadth excluding spines	14.3	13.8	13.0	16.2	16.1	18.1
Length of cheliped	83.0	87.0	78.0	117.0	110.0	107.0
Length of merus	33.0	36.0	32.0	49.0	46.0	45.0
Length of carpus	7.1	6.8	7.0	8.3	8.5	8.4
Length of chela	38.3	40.0	36.0	53.0	51.5	49.0
Length of dactylus	18.5	18.5	17.0	23.1	21.4	21.3
Length of first ambulatory leg	49.0	50.0	49.0	64.0	63.0	66.0
Greatest diameter of eye	4.3	4.2	4.0	4.6	4.5	4.7

The specimens from which the last four columns of measurements were taken have the chelæ modified, as usual in the old males of the species of the genus, by the proximal curvature and expansion of the digits, particularly the propodal, so as to leave them gaping at base; while the specimen from which the second column of measurements was taken has the chelæ slender and unmodified as in the female.

Specimens examined.

Station No.	Locality. N. lat.	Locality. W. long.	Depth in fathoms.	Nature of bottom.	When collected.	No. of specimens.	With eggs.
	° ′ ″	° ′ ″			1881.		
	OFF MARTHA'S VINEYARD.						
665	40 05 00	70 23 00	65	fne. S. M.	Sept. 4	2	
871	40 02 54	70 23 40	115	M. fne. S.	Sept. 4	150+	
872	40 05 30	70 23 52	86	S. G. Sh. & sponges.	Sept. 4	15	
873	40 02 00	70 57 00	100	sft. M.	Sept. 13	3	
874	40 00 00	70 57 00	85	sft. M.	Sept. 13	6	
877	39 56 00	70 54 18	126	sft. M.	Sept. 13	1	
878	39 55 00	70 54 15	142	M.	Sept. 13	6	
921	40 07 48	70 43 54	67	gn. M.	July 16	410+	⅞
922	40 03 48	70 45 54	71	gn. M. S.	July 16	1,300+	⅞
923	40 01 24	70 46 00	98	S.	July 16	8	
939	39 53 00	69 50 30	264	gn. M. S.	Aug. 4	1	
940	39 54 00	69 51 30	134	hrd. S. sponges.	Aug. 4	80+	⅞
941	40 01 00	69 56 00	79	hrd. S. M.	Aug. 4	500+	⅞
944	40 01 00	71 14 30	128	M. S. Sh.	Aug. 9	15	
949	40 03 00	70 31 00	100	rL M.	Aug. 23	500+	⅞
1028	39 58 00	70 06 00	146	S. and Sh.	Sept. 21	8	
1040	40 00 00	70 06 00	93	S. and Sh.	Sept. 21	7	
1151	39 58 30	70 37 00	125	S.	Oct. 4	1♂	
	OFF DELAWARE BAY.				1881.		
1043	38 39 00	73 11 00	130	S.	Oct. 10	154	18
1046	38 33 00	73 18 00	104	S.	Oct. 10	10	4
1047	38 31 00	73 21 00	156	S.	Oct. 10	2	
	OFF CHESAPEAKE BAY.				1880.		
896	37 26 00	74 19 00	56	S. Sh.	Nov. 16	3	
890	37 22 00	74 29 00	57	S.	Nov. 16	72	

The Blake dredgings of 1880 extend the range southward considerably beyond the above, as the following record of the occurrence of the species in these dredgings shows:

Station.	N. lat.			W. long.			Fathoms.	Specimens.
	o	′	″	o	′	″		
311	39	59	30	70	12	00	143	1
314	32	24	00	78	44	00	142	50+
315	32	18	20	78	43	00	225	1
333	35	45	25	74	50	30	65	100+
335	38	22	25	73	33	40	89	31
336	38	21	50	73	32	00	197	6
344	40	01	00	70	58	00	129	1

Munida valida, sp. nov.   (Pl. 1.)

A large species with the general appearance of *M. Bamffia*, but at once distinguished from it, and from *M. tenuimana*, and *Caribœa?* Smith as well, by the short and obtusely rounded epimera of all the abdominal somites.

Excluding the rostrum, the carapax is about three-fourths as broad as long; including the rostrum, about four-sevenths as broad as long, the rostrum being more than a fourth the entire length. The rostrum and the spines at its base are shorter and stouter than the *M. Bamffia*, and the latter are about three-fifths as long as the rostrum, strongly divergent and directed somewhat upward, while the rostrum is horizontal. The number and position of the spines on the dorsal surface and along the lateral margins of the carapax are very nearly as in *M. Bamffia*, except that there are no spines along the raised posterior margin. The orbital part of the anterior margin is more oblique than in *M. Bamffia*, and the antennal spine is not, as in that species, at the antero-lateral angle, but the margin between the antennal and hepatic spines is only a very little more oblique than the orbital margin, and the antero-lateral angle is really formed by the hepatic spine. The carapax is apparently wider and less convex than in *M. Bamffia*, the sutures of the dorsal surface are deeper, and the transverse rugæ are apparently fewer and more conspicuous.

The eyes are about as large as in *M. Bamffia*, but not so strongly compressed.

The basal segment of the antennula is armed with a slender spine arising from the prominence on the outer margin and directed forward, a larger spine on the outer edge of the distal end, and between these two a long spine, two-thirds as long as the segment itself, directed obliquely upward, while at the distal end of the inner side there is only an inconspicuous dentiform spine in place of the very long and slender spine found there in *M. Bamffia, tenuimana*, and *Caribœa?* Smith. The flagella of the antennæ are subcylindrical, slender, nearly naked, and not far from twice as long as the entire length of the body.

The merus of the external maxilliped is not distinctly tapered distally, and the ventral edge is armed with a slender spine at the distal end and a larger one a little way from the proximal end.

The chelipeds are equal, and in the male about two and a half times as long as the carapax, and resemble those of *M. Bamffia* very closely. In the male, the merus is nearly as long as the carapax, the carpus about two-fifths as long as the merus, and the chela much longer than the merus, much more slender, with the digits fully three-fourths as long as the body, slender, straight, and the prehensile edges in contact throughout. Although the single male seen is very large, there is no sign whatever of the expansion of the chela at the base of the digits, due largely to a curvature in the basal part of the propodal digit, which seems to be characteristic of the old males of all the species of the genus.

The dorsal surface of the abdomen is sculptured very much like the carapax, and the second and third somites are each armed with a series of small spines along the anterior edge above the facet, but there are no similar spines on the succeeding somites. The epimera of the second to the sixth somite are short, and obtusely rounded below, but those of the second and fifth are broader than the others. The telson and uropods are as in *M. Bamffia*.

As in all the other species of the genus which I have seen, the appendages of the first abdominal somite are shorter than those of the second, and composed of a slender protopod and a single thin lamella, which is much shorter than the protopod, broad, obtuse at the distal extremity, with a few marginal setæ, and rolled together anteriorly into a spoon-shaped appendage; while the protopod in the second pair of appendages is much longer than in the first, and bears a narrow, setigerous, and somewhat twisted lamella, with a minute rudiment of a second lamella at its base. The appendages of the third, fourth, and fifth somites are alike, and in each the protopod (apparently) is expanded into a broad oval lamella, margined with long setæ along the outer edge and at the tip, and bearing, on the inside near the tip, a small styliform appendage, composed of two segments. In the female the appendages of the second somite, though apparently not ovigerous, are about half as long as those of the third, with the protopod about as long as the endopod, which is composed of two subequal segments, and all the segments bear numerous long plumose setæ; the appendages of the third, fourth, and fifth somites are ovigerous, alike, nearly equal in size, and the two distal segments are subequal in length, and each about as long as the protopod.

I have seen only two specimens, from which the following measurements, in millimeters, were taken:

	1.	2.
Sex	♂	♀
Length, tip of rostrum to top of telson	83.0	70.0
Length of carapax, including rostrum	43.0	39.0
Length of rostrum	11.8	10.8
Breadth of carapax at cervical suture	20.0	18.0
Greatest breadth	24.0	22.0
Length of cheliped	110.0	75+
Length of merus	41.0	29.0
Length of carpus	16.0	14.0
Length of chela	48.0	28+
Length of dactylus	21.0	
Length of first ambulatory leg	77.0	62.0
Greatest diameter of eye	5.2	5.0
Length of telson	10.0	9.7
Breadth of telson	16.0	14.0

Station—	N. lat.	W. long.	Fathoms.	Specimens.
	° ′	° ′		
1112	39 56	70 35	243	1♂
1124	40 01	68 54	640	1♀

Eumunida, gen. nov.

The single species of the genus here proposed has the general appearance of *Munida*, but is at once distinguished from it and all the allied genera by the five-spined front, the position and structure of the peduncles of the antennæ, the absence of branchiæ at the bases of the external maxillipeds, the very broad and transversely segmented telson, and the absence of appendages upon the first five somites of the abdomen of the male.

The carapax is strongly contracted below anteriorly, so that the peduncles of the antennæ are near together and immediately beneath the well-developed eyes. The proximal segment of the peduncle of antennula is slender, subcylindrical, but with a small protuberance near the base where the auditory organ is situated, and unarmed. The peduncle of the antenna is highly developed and armed with numerous spines, of which one is articulated by a broad base to the second segment and evidently represents the antennal scale. The oral appendages and thoracic legs are similar to those of *Munida*, but there are neither branchiæ nor epipods at the bases of the external maxillipeds, though in other respects the branchial formula is the same. The telson is short and broad, more or less membranaceous, and divided by a transverse articulation, so that the distal part may be folded beneath the basal part. The female has well-developed appendages, all apparently origerous, upon the second to the fifth somite of the abdomen, but there are no appendages whatever on any of the first five somites in the adult male.

Eumunida picta, sp. nov. (Pl. 2, Fig. 2; Pl. 3, Figs. 6–10; Pl. 4, Figs. 1–3a.)

The carapax at the posterior part of the branchial region is about as

broad as the length, excluding the rostrum, but is rapidly narrowed anteriorly, and at the bases of the antennæ is scarcely half as broad. Back
of the cervical suture the dorsal surface is regularly convex transversely,
but the anterior part of the elevated gastric region is flat or slightly concave, and the orbital margins are perpendicular and hidden from above
by the bases of the supraorbital spines. The anterior edge of the front
is slightly arcuate and armed with five slender, acute, and subcylindrical spines, a median with two supraorbital each side; the median, or
rostrum proper, is about half as long as the rest of the carapax, straight
and horizontal; the supraorbital spines each side are approximately
parallel with the rostrum, but directed slightly upward so that their
tips are a little above the plane of the rostrum, are separated from the
rostrum more widely than from each other, and the inner is nearly
three-fourths as long as the rostrum while the outer is scarcely half as
long as the inner. Immediately back of the outer of these spines there
is a prominent and acute spine directed forward, and on a line between
this and the hepatic spine of the lateral margin there are two much
smaller spines on the steep side of the gastric region back of the orbit.
The lateral margin is arcuate in outline and armed with seven acute
spiniform teeth directed forward and decreasing successively in size
posteriorly; the anterior, or antennal, is separated from the base of the
antenna by a considerable space and is nearly as long as the outer
supraorbital spine, the second is on the hepatic region, and the remain-
ing five are all on the branchial region, the posterior one being very
small in adult specimens and nearly or quite obsolete in young specimens 15^{mm} in length. The dorsal surface is marked with transverse
rugæ, is sparsely clothed with minute hairs, and, except the spines
already mentioned, is unarmed. The cervical suture is well marked
and the gastro-hepatic distinct. The infero-lateral region is of nearly
the same form as in the typical species of *Munida* and terminates anteriorly in an acute spine a little in front of the first lateral spine.

The eyes are black, smaller than in the typical species of *Munida*,
nearly globular, and are borne on short stalks, the whole length being
scarcely more than a fourth greater than the diameter of the cornea.

The peduncle of the antennula reaches to about the tip of the rostrum; the segments are all approximately equal in length, nearly naked,
entirely unarmed, slender, and subcylindrical, though the proximal segment is considerably stouter than the others, and has a conspicuous
protuberance over the auditory organ. The upper flagellum is about
as long as the distal segment of the peduncle, swollen toward the base,
and tapered to a very slender tip. The lower flagellum is very slender
throughout and shorter than the upper. The peduncle of the antenna
reaches to about the tip of the second segment of the peduncle of the
antennula, and is armed with numerous spines; the first segment is exposed at the antero-lateral angle of the carapax and projects anteriorly
in a sharp tooth; the second segment is very short, armed externally

with a stout dentiform spine directed forward, and above bears a slender spiniform appendage curved slightly upward and outward, and a
little longer than the fourth segment; the third segment projects below
the fourth segment in a slender spiniform process reaching by the
fourth segment; the fourth segment is nearly as long as the diamete
of the eye, beyond which it reaches considerably, and is armed at the
distal end by a long spine projecting beneath and beyond the ultimate
segment, and above and on the outer side by two small teeth; the ultimate segment is little more than half as long as the fourth, about once
and a half as long as broad, and armed at the distal end with three
long and approximately equal and equidistant spines. The flagellum
is nearly as long as the whole body, slender, slightly compressed vertically, sparsely armed with minute setæ, and, at long intervals, with
a few very long and slender setæ.

The mandibles and maxillæ are very nearly as in *Munida Bamffia*,
but the proximal lobe of the protognath of the first maxilla is broader
and less prolonged and more obtusely rounded anteriorly.

The proximal lobe of the protopod of the first maxilliped projects
very little anteriorly, and the distal lobe is fully twice as long as broad.
The endopod projects considerably beyond the protopod, is less curved
than in *Munida Bamffia*, scarcely at all tapered distally, and clothed
with slender setæ along the inner edge and at the obtuse tip. The
basal portion of the exopod is longer than the endopod, from a sixth to
an eighth as broad as long, sparsely setigerous along the edges, and
bears a slender flagellum slightly less than half as long as the basal
part, and obscurely multiarticulate distally. The epipod is small, about
half as long as the endopod, tapered to the tip, and setigerous distally.

The second maxilliped resembles closely that of *Munida Bamffia*, but
the endopod is shorter and stouter, the merus being scarcely more than
twice as long as broad, and the basal part of the exopod is a little
shorter, scarcely narrowed distally, and somewhat less setigerous.

The ischium and merus in the external maxilliped are approximately
equal in length, the ischium unarmed at the distal end, but with the
inner angle dentate as usual; the merus is only very slightly expanded
on the inner side, and bears only a small spine near the distal end; the
propodus is narrow, with a very slight expansion on the inner side; and
the dactylus is considerably smaller than the propodus, and subcylindrical. The basal part of the exopod does not reach the distal end of
the merus. There are no maxillipedal arthrobranchiæ, as there are in
the species of *Munida*.

The chelipeds are not far from three times as long as the carapax,
including the rostrum, and are apparently not much shorter proportionally in the females and young than in the adult males. The merus
is subcylindrical, considerably longer than the carapax, including the
rostrum, and is armed with four longitudinal series of spines, of which
those forming the two series on the inner side are much larger than

those of the outer series, and these larger still than those of the lower series, which are quite small; there are eight to twelve of the larger spines in each series, and the surface between the spines, and also on the carpus and the body of the chela, is roughened with small squamiform and sparsely setigerous elevations. The carpus is short and armed with three distal spines on the inner side, and with a few small spines and tubercles on the outer side. The chela is just about as long as the merus and no stouter; the body is subcylindrical, considerably longer than the digits, and armed along the inner side with two series of spines corresponding with the two inner series on the merus, but the spines are much smaller and more crowded; the digits are slender, nearly straight laterally, but curved slightly downward at the tips, and the prehensile edges are irregularly dentate.

The first pair of ambulatory legs reach about to the middle of the carpi of the chelipeds; the dorsal edge of the merus is compressed and armed with a series of about ten large spines; the antero-inferior angle is armed with a similar series of much smaller spines, and there is, in addition, a large spine on the posterior side below the articulation with the carpus; the carpus is short and crested above with a series of spines like the merus, and the posterior side in both carpus and merus is roughened like the surface of the chelipeds; the propodus is about as long as the merus, slender, compressed laterally, with a few long setæ on the upper edge and a series of short spiniform setæ below, but without true spines or teeth; the dactylus is nearly half as long as the propodus, broad, strongly compressed, terminates in a strong chitinous tip, and is armed below with a closely set series of setiform chitinous spines decreasing in size proximally. The second pair are like the first, except that the merus is unarmed below. The third pair are considerably shorter than the second, reaching scarcely to the tips of the propodi of the second pair, and there is a series of small spines along the middle posterior side of the merus, but in other respects they are like the third pair.

The posterior pair of thoracic legs are much shorter than in the typical species of *Munida*, being only about as long as the meri of the third pair of ambulatory legs; the merus and carpus are about equal in length, and each is considerably longer than the ischium; the chela is little more than half as long as the carpus, but swollen distally, so as to be much broader, and the prehensile edge of the propodus and the articulation with short, stout, and strongly curved dactylus is terminal and nearly transverse, the propodal digit being reduced to a slight angular projection. The chela and distal end of the carpus are densely clothed with long setæ.

The consolidated sternal plates between the bases of the chelipeds and true ambulatory legs are marked by a deep longitudinal median sulcus on each somite, are separated from each other by conspicuous sulci, and the plate between the bases of the chelipeds is armed each side with a small spiniform tooth projecting forward, and the plane of the plate is much below the very narrow sternal plate at the bases of the external maxillipeds. The sternum of the last thoracic somite is entirely membranaceous, without any calcified plate or bar between the bases of the posterior legs.

The abdomen is broad, evenly rounded above, and without longitudinal carinæ; the epimera are all very short; and the sterna of all the somites are almost entirely membranaceous, like that of the last thoracic somite. The dorsum of the first somite rises in a sharp and very narrow transverse ridge back of the facet which slides beneath the carapax, and is inclosed either side by the anterior projection of the epimera of the second somite. The epimeron of the second somite is truncated below, but projects forward in a sharp angle at the side of the carapax, and above the angle is armed with a large, curved, and acute spine, directed forward above the lateral margin of the carapax. The epimera of the third, fourth, and fifth somites are truncated, with the angles more or less rounded, and those of the sixth obtuse. The second and third somites are each marked above by two transverse ciliated rugæ, the fourth and fifth each by three similar but less conspicuous rugæ in adults, or only two in the young, and the sixth somite and all the epimera are marked by broken and irregular rugæ or squamiform elevations. The sixth somite is much longer than the fifth, about a third as long as broad, and the postero-lateral edge outside the articulation of the uropod is oblique and nearly straight.

The telson in full-grown specimens is only as long as the sixth somite, and twice as broad as long, but in young specimens is proportionally longer and narrower. The whole appendage is thin and slightly calcified; the lateral margins are deeply incised about the middle and the incisions connected by a transverse membranous articulation, so that the distal part is readily folded beneath the proximal. The distal part is notched at the middle of the posterior edge and longitudinally divided by a membranous line, so that it appears to be formed of two transverse elliptical plates, each nearly twice as broad as long, and of which the posterior and lateral edges are thickly ciliated. The inner lamella of the uropod is fully as long as the telson, about two-thirds as broad as long, elliptical, the inner and distal edges armed with spines, which are small on the inner and very minute and crowded on the distal edge, and the entire margin, except near the base, is ciliated with numerous long hairs. The outer lamella is longer and broader than the inner, narrowed and somewhat excavated on the inner edge near the base, and margined with hairs like the inner.

There are no appendages whatever on any of the first five abdominal somites in any of the adult males examined. In young specimens, 15^{mm} or less in length, in which the sexual characters are not manifest, but which are possibly immature males, or more probably immature females, there are, however, on the second to the fifth somite, rudimentary, very

Vol. VI, No. 4.   Washington, D. C.   June 20, 1883.

minute, and naked appendages, obscurely divided into a large proximal and a small distal segment. In the adult female the appendages of the second to the fifth somite are similar, approximately alike in size, apparently all ovigerous, and each appears to be composed of only two segments, of which the distal is about half as long as the proximal. None of the specimens seen are carrying eggs.

Five specimens give the following measurements in millimeters:

	Station—				
	1152.	1152.	1043.	1097.	1097.
Sex	Young.	♂	♂	♀	♀
Length, tip of rostrum to tip of telson	15.0	23.5	43.0	24.0	40.0
Length of carapax, including rostrum	8.5	13.0	23.7	13.4	22.1
Length of rostrum	2.5	4.5	8.1	4.6	8.0
Breadth at bases of antennal spines	4.1	6.0	11.0	6.2	10.2
Greatest breadth, including spines	5.5	8.5	15.6	9.0	14.5
Length of cheliped	27.0	34.0	68.0	40.0	57.0
Length of merus	10.5	15.0	30.0	18.0	24.0
Length of carpus	1.5	2.5	5.2	2.6	4.3
Length of chela	11.0	15.0	30.0	18.0	24.5
Length of dactylus	4.7	6.4	14.0	7.5	11.4
Length of first ambulatory leg	13.5	20.0	38.0	21.0	35.0
Diameter of eye	1.3	1.7	3.0	1.8	3.0
Length of telson	1.4	2.1	4.0	2.2	4.0
Breadth of telson	2.0	3.7	8.0	3.9	8.0

Specimens examined.

Station No.	Locality.		Depth in fathoms.	Nature of bottom.	When collected.	No. of specimens.		With eggs.
	N. lat.	W. long.				♂	♀	
	OFF MARTHA'S VINEYARD.							
1038	° ′ 39 58	° ′ 70 06	146	S. Sh.	1881. Sept. 21	1 1 y. 1	0	
1097	39 54	69 44	158	fine S.	1882. Aug. 11	1	2	0
1098	39 53	69 43	156	fine S.	Aug. 11	1	1	0
1152	39 58	70 35	115	S.	Oct. 4	2 1 y.		
	OFF DELAWARE BAY.							
1043	38 39	73 11	130	S.	1881. Oct. 10	1		

In the specimen from station 1152, after preservation in alcohol for a short time, the coloration had apparently not changed very materially, and was very striking. The whole dorsal surface of the carapax and abdomen was light red, lightest on the abdomen and darkest on the rostrum and spines of the carapax. The chelipeds and three pairs of ambulatory legs were very intense bright red, except the digits of the chelæ and the distal extremities of the ambulatory legs to very near the bases of the propodi, which parts were white, the color stopping on each of these appendages very suddenly at the point where they cease to be armed with spines. All the other specimens show more or less distinct indications of the same coloration.

Anoplonotus politus, gen. et sp. nov. (Pl. 2, Fig. 1; Pl. 3, Figs. 1–5 a.)

Excluding the rostrum, the carapax is nearly as broad as long; including the rostrum, seven to eight tenths as broad as long, the rostrum being rather less than a fourth of the entire length. The rostrum is vertically flattened, though obscurely carinated longitudinally above, horizontally triangular, but not acute at the tip, and slightly curved downward distally. There are no spines or tubercles upon the carapax, but the gastric region is somewhat protuberant and separated from the branchial regions by a broad sulcus each side, and from the prominent cardiac region by a still deeper sulcus which extends either side as a shallow sulcus across the branchial region, which is again crossed by a narrower sulcus in front, but the cardiac region is not conspicuously separated from the branchial region either side of it. The orbital portion of the anterior margin is narrow and advanced considerably in front of the antero lateral angles, which are formed by the hepatic regions and are nearly right-angular. The lateral margins are slightly curved, and the greatest breadth is a little back of the middle. The surface of the carapax is granulose, particularly along the sides, where the granules are arranged in transverse lines.

The small eyes are partially beneath the rostrum, and scarcely reach its middle; there is a slight protuberance on the outer side of the stalk near the base; and the eye itself is semitranslucent in the alcoholic specimens; its diameter is rather less than that of the stalk and about half the whole length, and the cornea is apparently entirely without facets.

The basal segment of the peduncle of the antennula is a little shorter than the rostrum, about three-fourths as broad as long, somewhat swollen on the outer side, and armed with two teeth at the distal extremity. The second and third segments are slender, subequal in length and each scarcely as long as the basal. The upper flagellum is about as long as the distal segment of the peduncle, the basal portion swollen and composed of numerous short segments, while the distal portion is very slender and composed of about five elongated segments. The lower flagellum is little more than half as long as the upper, slender throughout, and composed of about three segments. The peduncle of the antenna arises just outside the peduncle of the antennula and at some distance from the antero-lateral angle of the carapax, scarcely reaches the tip of the rostrum, and its three distal segments are slender. The flagellum is very slender, and reaches to about the tips of the chelipeds.

The mandibles are of essentially the same form as in *Munida;* the molar area is transverse to the body of the mandible, narrow, naked, and separated from the broad and edentulous ventral process by a deep excavation; and the palpus is slender, triarticulate, and armed with few and short setæ. The protognathal lobes of the first maxilla are approximately equal in size, rather broad at the ends, and armed as usual with slender spines upon the distal, and numerous setæ upon the proximal lobe. The endognath is much shorter than the distal lobe of the protognath, slender, tapered to an obtuse point, and armed with two series of small setæ, one at the tip and the other below the middle. The protognathal lobes of the second maxilla are approximately equal in size, and each lobe is divided into two lobules very unequal in width, the two middle lobules being approximately a third as wide as the anterior and posterior, though all four of the lobules are of about the same length. The endognath is a little longer than the distal lobe of the protognath, tapers to a slender tip, and is armed with a very few setæ along the middle of its length. The anterior portion of the scaphognath is a little shorter than the endognath, broad, slightly narrowed anteriorly, but broad and obtusely rounded at the tip, while the posterior portion is short, transversely truncated behind, broader than long, and somewhat triangular in outline, with the angles rounded.

The tips of the lobes of the protopod of the first maxilliped are rounded and nearly alike, but the distal lobe is considerably longer than the proximal, being about twice as long as broad. The endopod is about as long as the distal lobe of the protopod, narrow, tapered to an obtuse tip, very strongly curved, the outer edge margined with slender setæ distally, and proximally with a very few setæ near the inner edge. The exopod is lamellar, a little longer than the endopod and much broader, being about a fifth as broad as long, rapidly narrowed at the extremity, and margined with slender setæ along the outer edge. The epipod is about as broad but scarcely as long as the distal lobe of the protopod, triangular at the extremity, and ciliated at the tip and along the outer edge. The endopod of the second maxilliped is of nearly equal breadth from the base to the dactylus; the ischium is scarcely longer than broad; the merus nearly three times as long as broad and about as long as the three terminal segments taken together; the carpus and propodus are subequal in length; the dactylus is shorter and much narrower than the propodus, and rounded at the tip; and all the segments are more or less armed as usual with setæ of different forms, and at the distal end of the inner edge there is a single slender spine or spiniform setæ in addition to a few short setæ. The exopod is much larger than the endopod; the unsegmented basal portion is nearly uniform in breadth for about the proximal two-thirds of its length, where it expands in an obtuse prominence opposite the carpus of the endopod, but from this prominence it tapers to the articulation with the slender flagellum; except near the tip both edges are margined with short setæ, and the prominence of the inner edge bears in addition a submarginal series of six to eight long setæ; the flagellum is about half as long as the basal part, distinctly articulated near the middle, and the terminal fourth of the whole length very obscurely multiarticulate and furnished with long setæ.

The external maxillipeds, when extended, reach a little by the tip of the rostrum; the ischium is nearly twice as long as broad and triquetral, with the dorso-internal angle sharply and regularly dentate; the merus is slightly shorter than the ischium, nearly as broad as long, expanded distally and armed with two obtuse teeth on the inner side and with a single tooth outside the articulation of the carpus; the three distal segments are slender and together about equal in length to the ischium and merus, the propodus being about as long as the merus, and the carpus and dactylus successively a little shorter. The epipod and exopod are well developed, and the basal part of latter reaches considerably by the merus.

The chelipeds are equal and about three times as long as the carapax, the merus being about as long as the carapax, and the chela considerably longer. In the females and young males the chelipeds are very slender and subcylindrical throughout, with the chela scarcely, if at all, stouter than the carpus, and with the digits straight and very slender. In the large males the chelipeds are very much stouter; the body of the chela is expanded and vertically flattened distally, and the digits gape widely at the base, the proximal half of the propodal digit being strongly curved and unarmed, while the distal part of the prehensile edge is straight and minutely serrate like the corresponding part of the dactylus, with which it is in contact when the digits are closed; the basal part of the dactylus is only slightly curved, but is armed with an obtuse tubercle on the prehensile edge near the base; and the whole prehensile edges of both digits are more or less hairy.

The three pairs of ambulatory legs are slender and subequal in length, about as long as the body, and the dactyli are slender, strongly curved, more than half as long as the propodi, and unarmed. The posterior legs are very small and slender and of essentially the same form as in *Munida*. There are no epipods at the bases of any of the thoracic legs.

The abdomen is considerably shorter and much narrower than the carapax, and its dorsal surface is nearly smooth and devoid of carinæ, except on the edges of the sixth somite, as described beyond, though there is a slight transverse sulcus on the middle portion of the second somite, which is also raised sharply above the small facet which slides under the carapax. The epimera of the second somite are broad; the third and fourth somites are short, and their epimera very narrow and acute; the fifth somite is a little longer than the fourth, and its epimera broader and more obtuse than those of the fourth; while the sixth somite is slightly longer than the fifth, and the postero-lateral margins

of its epimera are excavated to fit accurately the outer edges of the
bases of the uropods, and are margined with a narrow carina.

The telson is approximately two-thirds as long as broad, narrowed
posteriorly, with the posterior angles rounded and the posterior edge
slightly emarginate in the middle. The telson is stiffened by eight
distinct calcified plates; a broad median basal plate, with a small one
either side at the base of the uropod and a small median one behind it
and between a pair of broad lateral plates, still behind which there is a
second pair which meet in the middle line and form the tips and lateral
angles.

The lamellæ of the uropods are about as long as the telson, a little
longer than wide, the inner slightly longer than the outer, and each
widest near the extremity, which is broadly rounded in outline, while
the outer edge is nearly straight.

In the male the first and second pairs of abdominal appendages are
well developed and of nearly the same form as in the species of *Munida*.
Those of the first pair are about as long as the protopods of the second
pair, with the protopod somewhat triquetral and naked except a few
setæ along the distal part of the inner edge, and with the single terminal lamella slightly shorter than the protopod but much broader, very
thin, margined with setæ distally and along the outer edge, and the
edges rolled together on the anterior side. In the second pair the protopod is slender and naked, and bears a narrow, lanceolate lamella a
little shorter than the protopod and clothed with numerous setæ along
both edges and on the proximal part of the anterior side, and at its
base a minute second lamella much narrower than the other, scarcely
as broad as long, and naked. The appendages of the third, fourth, and
fifth somites are rudimentary, very minute, and almost wholly naked;
they are scarcely an eighth as long as the appendages of the second
somite, very slender, and with a single terminal lamella smaller than
the protopod. In the female there are no appendages upon the first
somite of the abdomen, and the appendages of the second somite are
very minute, slender, and tipped with a few small setæ. The appendages of the third, fourth, and fifth somites are well developed, unira-
mous, and ovigerous; they increase in size successively from the third to
the fifth, and each appendage is composed of a slender protopod and a
shorter terminal portion composed of two segments of which the terminal is the longer.

The eggs in the alcoholic specimens are approximately spherical,
1.50^{mm} to 1.75^{mm} in diameter, and very few in number, the two largest
egg-bearing specimens carrying less than thirty eggs each, while the
three smaller specimens carry nine, three, and two each, though a very
few eggs may have been lost from these last specimens.

Three specimens from station 941 give the following measurements in millimeters:

	1.	2.	3.
Sex	♂	♂	♀
Length, tip of rostrum to tip of telson	22.0	17.5	17.5
Length of carapax, including rostrum	12.2	9.4	9.3
Length of rostrum	3.0	2.3	2.2
Breadth of carapax at anterior angles	7.0	5.4	5.3
Greatest breadth of carapax	9.0	7.3	7.0
Length of cheliped	35.0	31.0	28.0
Length of merus	12.0	11.0	10.0
Length of carpus	5.0	4.3	3.8
Length of chela	15.9	13.3	11.5
Greatest breadth of chela	4.0	2.1	1.3
Length of dactylus	6.5	5.2	4.6
Length of first ambulatory leg	22.0	19.0	17.5
Length of telson	3.3	2.5	2.5
Breadth of telson	4.5	3.6	3.5
Diameter of eye	0.6	0.5	0.5

Specimens examined.

Station No.	Locality.						Depth in fathoms.	Nature of bottom.	When collected.	No. of specimens.		With eggs.	Dry or alcoholic.
	N. lat.			W. long.						♂	♀		
	°	′	″	°	′	″							
	OFF MARTHA'S VINEYARD.												
871	40	02	54	70	23	40	115	fne. S. M.	1880. Sept. 4		1	0	Alc.
873	40	02	00	70	57	00	100	sft. M.	Sept. 13	1	1	1	Do.
874	40	00	00	70	57	00	85	M.	Sept. 13		1	0	Do.
940	39	54	00	69	51	30	134	hrd. S. and sponges.	1881. Aug. 4	2	1	1	Do.
941	40	01	00	69	56	00	79	hrd. S., M.	Aug. 4	13	3	3	Do.

In the manuscript sent to the printer I referred this species to *Elasmonotus* with considerable hesitation, though it agreed very well with the brief diagnosis given by Milne-Edwards (Bull. Mus. Comp. Zool. Cambridge, viii, p. 60). The recently published figures of *E. Vaillantii* (Recueil de figures de Crustacés nouveau ou peu connus, April, 1883), however, seem to show that my species is generically as well as specifically distinct from *Elasmonotus*, being distinguished by the short and broad merus of the external maxilliped, the absence of spines, teeth, or carinæ upon the carapax and abdomen, and by the greater breadth of the carapax, if the measurements given by Milne-Edwards, are correct.* The species here described is apparently also distinguished generically by the small and non-segmented exopod of the first maxilliped, and specially by the rudimentary character of the appendages of the third, fourth, and fifth somites of the abdomen. The number and arrangement of the branchiæ, as indicated in the following formula, is the same as in *Munida:*

	Somite—								
	VII.	VIII.	IX.	X.	XI.	XII.	XIII.	XIV.	Total.
Epipods	1	0	1	0	0	0	0	0	(2)
Podobranchiæ	0	0	0	0	0	0	0	0	0
Arthrobranchiæ	0	0	2	2	2	2	2	0	10
Pleurobranchiæ	0	0	0	0	1	1	1	1	4
									14+(2)

*There is a perplexing disagreement in Milne-Edwards's characterization of his species between the descriptions of the proportions of the carapax and the accompanying measurements. *E. brevimanus* and *abdominalis* are each said to have the carapax narrower ("plus étroite") than *E. longimanus*, though the measurements given show *E. brevimanus* to be very much, and *E. abdominalis* slightly, broader than *E. longimanus.*

NEW HAVEN, CONN., *December 28, 1882.*

EXPLANATION OF PLATES.

All the figures on Plates I and II; Figs. 4 and 5, Plate IV; Figs. 4, 4a, 4b, and 5, Plate V; and Fig. 5, Plate VI, were drawn by J. H. Emerton. All the other figures were drawn by the author.

PLATE I.

Munida valida Smith. Dorsal view of male, from station 1112, natural size.

PLATE II.

FIG. 1.—*Anoplonotus politus* Smith. Dorsal view of a male, from station 941, enlarged two diameters.

FIG. 2.—*Eumunida picta* Smith. Dorsal view of a male, from station 1043, natural size.

PLATE III.

FIG. 1.—*Anoplonotus politus.* First maxilla of the right side, seen from below, of a male from station 941, enlarged twelve diameters.

FIG. 2.—Second maxilla of the right side of the same specimen, enlarged twelve diameters.

FIG. 3.—First maxilliped of the right side of the same specimen, enlarged twelve diameters.

FIG. 4.—Second maxilliped of the right side of the same specimen, enlarged twelve diameters.

FIG. 5.—External maxilliped of the right side of same specimen, enlarged eight diameters.

FIG. 5a.—Ischium and merus of the same appendage, seen from above, enlarged eight diameters.

FIG. 6.—*Eumida picta.* First maxilla of the right side of a male, from station 1098, seen from below, enlarged eight diameters.

FIG. 7.—Second maxilla of the right side of the same specimen, enlarged eight diameters.

FIG. 8.—First maxilliped of the right side of the same specimen, enlarged eight diameters.

FIG. 9.—Posterior thoracic leg of the same specimen, enlarged eight diameters.

FIG. 10.—Appendage of the fifth somite of the abdomen of a young specimen, 15^{mm} long, from station 1152, enlarged twenty-four diameters.

FIG. 11.—*Munida Caribæa ?* Smith. First maxilliped of a male, from station 1043, enlarged eight diameters.

PLATE IV.

FIG. 1.—*Eumida picta.* Extremity of the abdomen of a male, from station 1098, dorsal view, enlarged three and a half diameters.

FIG. 2.—Extremity of the abdomen of a young male, from station 1152, enlarged four diameters.

FIG. 3.—Peduncle of right antenna of a male, dorsal view, from station 1152, enlarged eight diameters; *a*, acicle, or articulated spine, of the second segment, representing the antennal scale; *b*, third segment, projecting anteriorly in a long spine.

Fig. 3a.—The same, side view; *a*, as in last figure.
Fig. 4.—*Eupagurus politus* Smith. Lateral view of left side of a male, from station 922, natural size.
Fig. 5.—*Catapagurus Sharreri* A. M.-Edwards. Lateral view of left side of a male in a carcinœcium, formed by *Adamsia sociabilis* Verrill, from station 940, enlarged two diameters.

Plate V.

Fig. 1.—*Eupagurus bernhardus* Brandt. Outline of transverse section through the lower part of the anterior arthrobranchia of the thirteenth somite (penultimate thoracic), showing the form of the lamellæ, enlarged eight diameters; *a*, afferent, and *b*, efferent vessel.
Fig. 2.—*Sympagurus pictus* Smith. Outline of similar section of the corresponding branchia of a female, from station 924, enlarged eight diameters, and lettered as in the last figure.
Fig. 2a.—Extremity of the same branchia, side view, enlarged eight diameters.
Fig. 3.—*Parapagurus pilosimanus* Smith. Outline of similar section of the corresponding branchia of a male, from station 880, enlarged eight diameters, and lettered as in Figs. 1 and 2.
Fig. 3a.—Extremity of the same branchia, side view, enlarged eight diameters.
Fig. 4.—*Parapagurus pilosimanus*. Lateral view of the left side of the originally described male specimen, taken on a trawl-line off Nova Scotia, half natural size.
Fig. 4a.—Dorsal view of the carapax and anterior appendages of the same specimen, natural size.
Fig. 4b.—Dorsal view of the chelipeds of the same specimen, half natural size.
Fig. 5.—Dorsal view of a male in the carcinœcium (*Epizoanthus paguriphilus* Verrill), from station 947, natural size.

Plate VI.

Fig. 1.—*Parapagurus pilosimanus*. First maxilla of the right side, seen from below, of a male from station 880, enlarged six diameters.
Fig. 2.—Second maxilla of the right side of the same specimen, enlarged six diameters.
Fig. 3.—First maxilliped of the right side of the same specimen, enlarged six diameters.
Fig. 4.—Appendage of the right side of the first somite of the abdomen of the same specimen, seen from behind, enlarged four diameters.
Fig. 4a.—Appendage of the right side of the second somite of the abdomen of the same specimen, seen from behind, enlarged four diameters.
Fig. 5.—*Sympagurus pictus*. Dorsal view, from life, of a male in the carcinœcium (*Urticina consors* Verrill), from station 924, one-half natural size.
Fig. 6.—First maxilla of the right side of a female, from station 1114, enlarged six diameters.
Fig. 7.—Second maxilla of the right side of the same specimen, enlarged six diameters.
Fig. 8.—First maxilliped of the right side of the same specimen, enlarged six diameters.
Fig. 9.—*Eupagurus bernhardus*. First maxilliped of the right side of a male, from station 119 (Halifax, Nova Scotia), enlarged six diameters.

532

PLATE I.

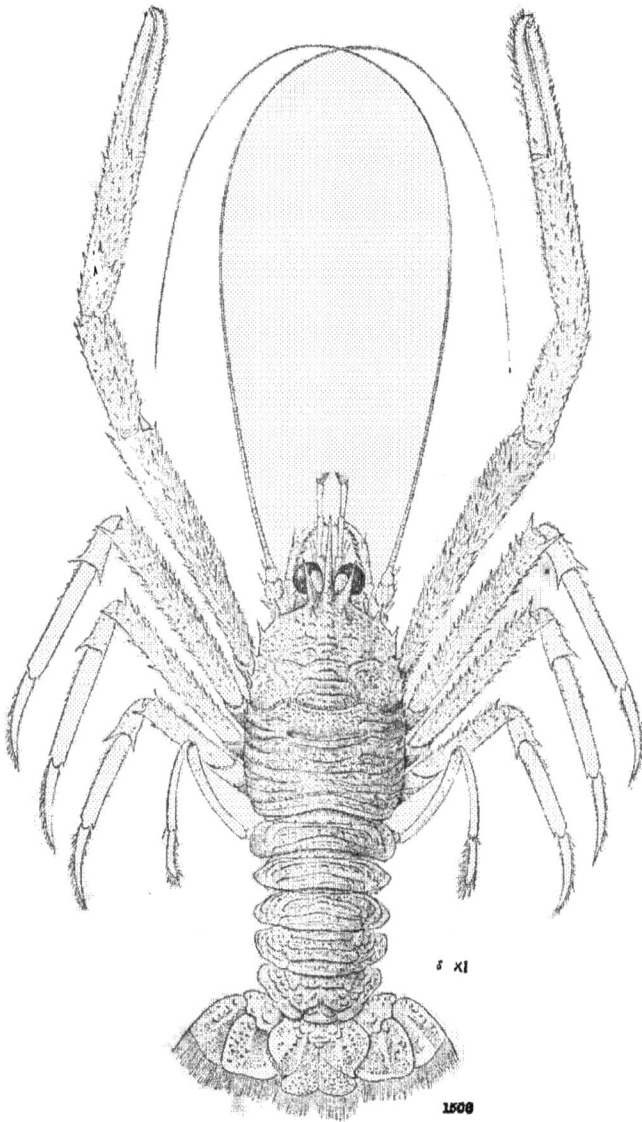

♂ ×1

1508

· 534

PLATE II.

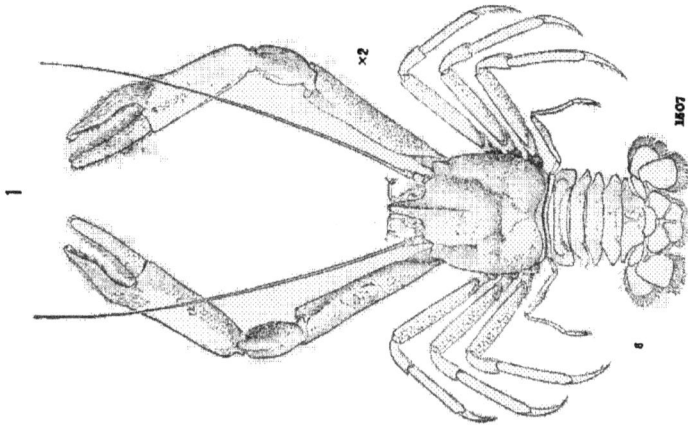

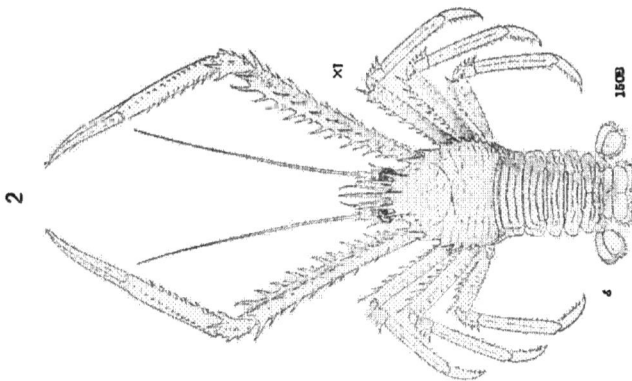

536

PLATE III.

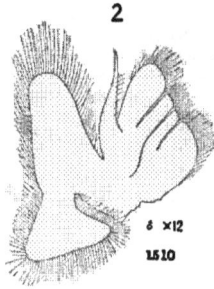

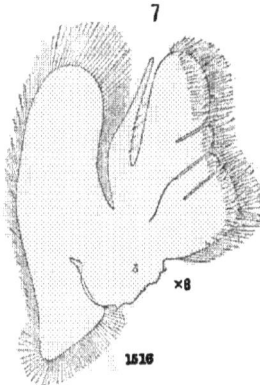

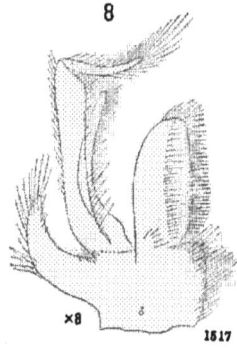

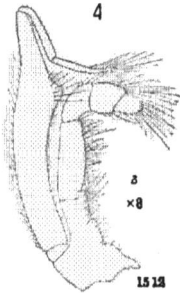

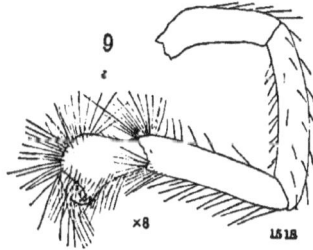

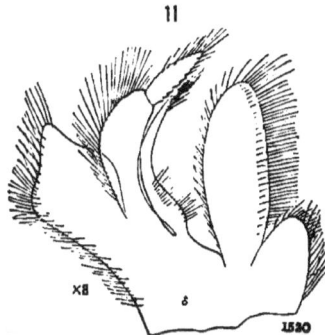

PLATE IV.

(Drawings of Figs. 4 and 5, by Mr. J. H. Emerton; the rest by Prof. S. I. Smith.)

FIG. 1—*Eumunida picta.* (p. 44.) Extremity of the abdomen of a male, from station 1006, dorsal view, enlarged three and a half diameters.

FIG. 2.—Extremity of the abdomen of a young male, from station 1152, enlarged four diameters.

FIG. 3.—Peduncle of right antenna of a male, dorsal view, from station 1152, enlarged eight diameters; a, acicle, or articulated spine, of the second segment, representing the antennal scale. b, third segment, projecting anteriorly in a long spine.

FIG. 3a.—The same, side view; a, as in last figure.

FIG. 4.—*Eupagurus politus* Smith. (p. 27.) Lateral view of left side of a male, from station 922, natural size.

FIG. 5.—*Catapagurus Sharreri* A. M.-Edwards. (p. 31.) Lateral view of left side of a male in a carcinœcium, formed by *Adamsia sociabilis* Verrill; from station 940, enlarged two diameters.

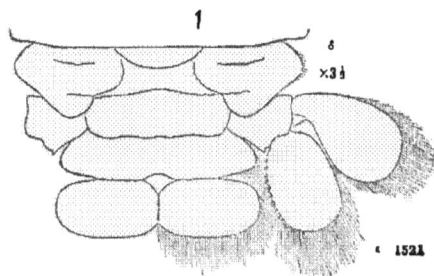

1

δ
×3½

1521

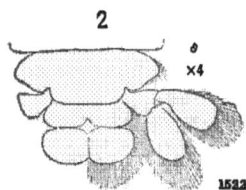

2

δ
×4

1522

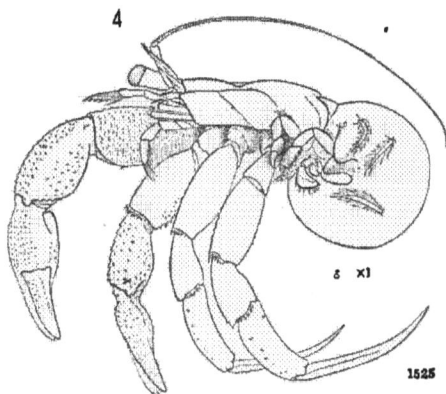

4

δ ×1

1525

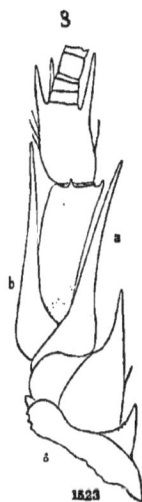

3

a

b

δ

1523

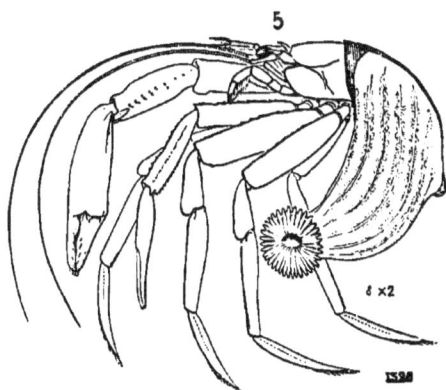

5

δ ×2

1526

3a

a

1524

PLATE V.

(Drawings of Figs. 4, 4a, 4b, and 5, by Mr. J. H. Emerton; the rest by Prof. S. I. Smith.)

FIG. 1.—*Eupagurus bernhardus* Brandt. (pp. 28, 29, *et seq.*) Outline of transverse section through the lower part of the anterior arthrobranchia of the thirteenth somite (penultimate thoracic), showing the form of the lamellæ; enlarged eight diameters; *a*, afferent, and *b*, efferent vessel.

FIG. 2.—*Sympagurus pictus* Smith. (p. 37.) Outline of similar section of the corresponding branchia of a female, from station 924, enlarged eight diameters, and lettered as in the last figure.

FIG. 2a.—Extremity of the same branchia, side view, enlarged eight diameters.

FIG. 3.—*Parapagurus pilosimanus* Smith. (p. 33.) Outline of similar section of the corresponding branchia of a male, from station 880, enlarged eight diameters, and lettered as in Figs. 1 and 2.

FIG. 3a.—Extremity of the same branchia, side view, enlarged eight diameters.

FIG. 4.—*Parapagurus pilosimanus.* (p. 33.) Lateral view of the left side of the originally described male specimen, taken on a trawl-line off Nova Scotia, half natural size.

FIG. 4a.—Dorsal view of the carapax and anterior appendages of the same specimen. natural size.

FIG. 4b.—Dorsal view of the chelipeds of the same specimen, half natural size.

FIG. 5.—Dorsal view of a male in the carcinœcium (*Epizoanthus paguriphilus* Verrill), from station 947, natural size.

540

PLATE V.

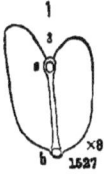

1

1527

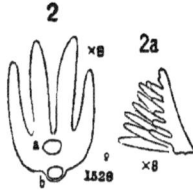

2

2a

1528

3

3a

1529

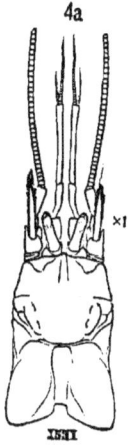

4a

1531

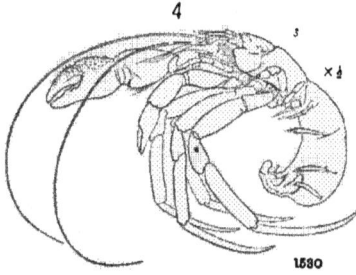

4

1530

4b

1532

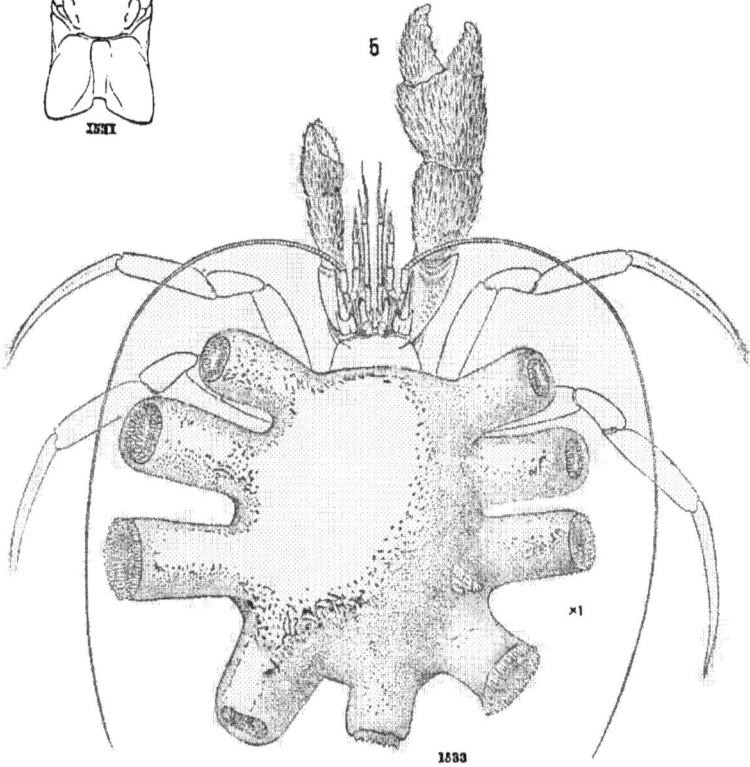

5

1533

PLATE VI.

(Drawing of Fig. 5, by Mr. J. H. Emerton; the rest by Prof. S. I. Smith.)

FIG. 1.—*Parapagurus pilosimanus.* (p. 33.) First maxilla of the right side, seen from below, of a male from station 880, enlarged six diameters.

FIG. 2.—Second maxilla of the right side of the same specimen, enlarged six diameters.

FIG. 3.—First maxilliped of the right side of the same specimen, enlarged six diameters.

FIG. 4.—Appendage of the right side of the first somite of the abdomen of the same specimen, seen from behind, enlarged four diameters.

FIG. 4a.—Appendage of the right side of the second somite of the abdomen of the same specimen, seen from behind, enlarged four diameters.

FIG. 5.—*Sympagurus pictus.* (p. 37.) Dorsal view, from life, of a male in the carcinœcium (*Urticina consors* Verrill), from station 924, one-half natural size.

FIG. 6.—First maxilla of the right side of a female, from station 1114, enlarged six diameters.

FIG. 7.—Second maxilla of the right side of the same specimen, enlarged six diameters.

FIG. 8.—First maxilliped of the right side of the same specimen, enlarged six diameters.

FIG. 9.—*Eupagurus bernhardus.* (pp. 28, 29, *et seq.*) First maxilliped of the right side of a male, from station 119 (Halifax, Nova Scotia), enlarged six diameters.

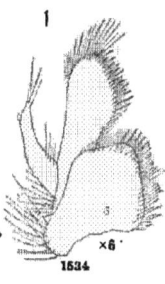

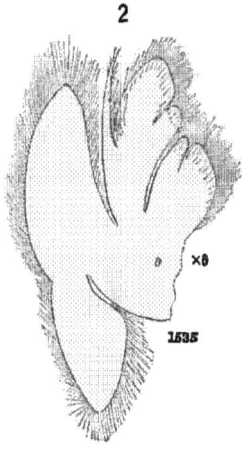

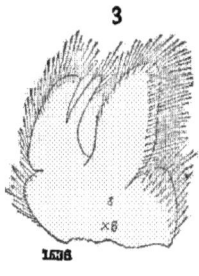

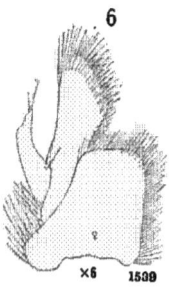

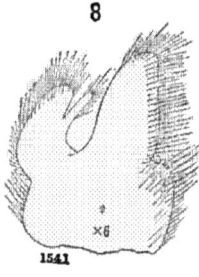

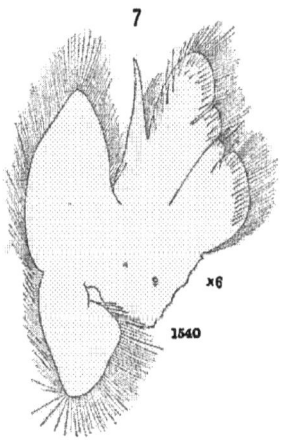

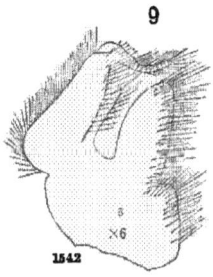

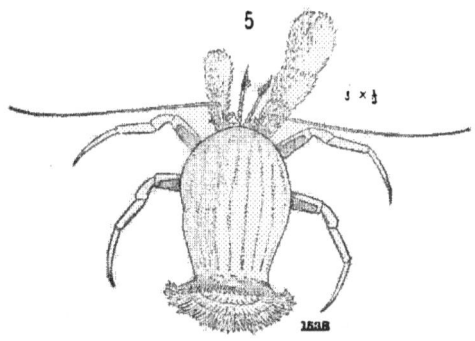